AF358677

Age-Related Musculoskeletal Disorders

Age-Related Musculoskeletal Disorders

Editors

Milica Lazovic
Dejan Nikolić

Basel • Beijing • Wuhan • Barcelona • Belgrade • Novi Sad • Cluj • Manchester

Editors
Milica Lazovic
State University of Novi
Pazar
Novi Pazar
Serbia

Dejan Nikolić
University of Belgrade
Belgrade
Serbia

Editorial Office
MDPI
St. Alban-Anlage 66
4052 Basel, Switzerland

This is a reprint of articles from the Special Issue published online in the open access journal *Medicina* (ISSN 1648-9144) (available at: https://www.mdpi.com/journal/medicina/special_issues/M6Y6965L6M).

For citation purposes, cite each article independently as indicated on the article page online and as indicated below:

Lastname, A.A.; Lastname, B.B. Article Title. *Journal Name* **Year**, *Volume Number*, Page Range.

ISBN 978-3-7258-0281-4 (Hbk)
ISBN 978-3-7258-0282-1 (PDF)
doi.org/10.3390/books978-3-7258-0282-1

Contents

Article

Combined Rehabilitation with Alpha Lipoic Acid, Acetyl-L-Carnitine, Resveratrol, and Cholecalciferolin Discogenic Sciatica in Young People: A Randomized Clinical Trial

Dalila Scaturro [1,*], Fabio Vitagliani [2], Sofia Tomasello [3], Cristiano Sconza [4], Stefano Respizzi [4] and Giulia Letizia Mauro [1]

[1] Department of Surgical, Oncological and Stomatological Disciplines, University of Palermo, 90144 Palermo, Italy; giulia.letiziamauro@unipa.it

[2] Department of Biomedical and Biotechnological Sciences, University of Catania, 95123 Catania, Italy; fabiovitagliani93@libero.it

[3] Faculty of Medicine and Surgery, University of Palermo, 90127 Palermo, Italy; sofiatomasello@alice.it

[4] Department of Physical and Rehabilitation Medicine, Humanitas Clinical and Research Centre IRCSS, 20019 Milan, Italy; cristiano.sconza@humanitas.it (C.S.); stefano.respizzi@humanitas.it (S.R.)

* Correspondence: dalila.scaturro@unipa.it; Tel.: +39-320-694-5411

Abstract: *Background and Objectives*: In the Western world, back pain and sciatica are among the main causes of disability and absence from work with significant personal, social, and economic costs. This prospective observational study aims to evaluate the effectiveness of a rehabilitation program combined with the administration of Alpha Lipoic Acid, Acetyl-L-Carnitine, Resveratrol, and Cholecalciferol in the treatment of sciatica due to herniated discs in young patients in terms of pain resolution, postural alterations, taking painkillers, and quality of life. *Materials and Methods*: A prospective observational study was conducted on 128 patients with sciatica. We divided the sample into 3 groups: the Combo group, which received a combination of rehabilitation protocol and daily therapy with 600 mg Alpha Lipoic Acid, 1000 mg Acetyl-L-Carnitine, 50 mg Resveratrol, and 800 UI Cholecalciferol for 30 days; the Reha group, which received only a rehabilitation protocol; and the Supplement group, which received only oral supplementation with 600 mg Alpha Lipoic Acid, 1000 mg Acetyl-L-Carnitine, 50 mg Resveratrol, and 800 UI Cholecalciferol. Clinical assessments were made at the time of recruitment (T0), 30 days after the start of treatment (T1), and 60 days after the end of treatment (T2). The rating scales were as follows: the Numeric Rating Scale (NRS); the Oswestry Disability Questionnaire (ODQ); and the 36-item Short Form Health Survey (SF-36). All patients also underwent an instrumental stabilometric evaluation. *Results*: At T1, the Combo group showed statistically superior results compared to the other groups for pain ($p < 0.05$), disability ($p < 0.05$), and quality of life ($p < 0.05$). At T2, the Combo group showed statistically superior results compared to the other groups only for pain ($p < 0.05$) and quality of life ($p < 0.05$). From the analysis of the stabilometric evaluation data, we only observed a statistically significant improvement at T2 in the Combo group for the average X ($p < 0.05$) compared to the other groups. *Conclusions*: The combined treatment of rehabilitation and supplements with anti-inflammatory, pain-relieving, and antioxidant action is effective in the treatment of sciatica and can be useful in improving postural stability.

Keywords: sciatica; rehabilitation; postural balance; alpha lipoic acid; acetyl-l-carnitine

Citation: Scaturro, D.; Vitagliani, F.; Tomasello, S.; Sconza, C.; Respizzi, S.; Letizia Mauro, G. Combined Rehabilitation with Alpha Lipoic Acid, Acetyl-L-Carnitine, Resveratrol, and Cholecalciferolin Discogenic Sciatica in Young People: A Randomized Clinical Trial. *Medicina* **2023**, *59*, 2197. https://doi.org/10.3390/medicina59122197

Academic Editors: Dejan Nikolić and Milica Lazovic

Received: 5 November 2023
Revised: 11 December 2023
Accepted: 14 December 2023
Published: 18 December 2023

1. Introduction

The most common cause of low back pain (LBP) and sciatica is represented by lumbar disc herniation (LDH) [1].

About 9% of all people in the world are affected. It has a substantial impact on quality of life and represents a significant economic burden. In the Western world, back pain and sciatica are among the main causes of disability and absence from work with significant personal, social, and economic costs. In recent years, lifestyle changes have led to a gradual increase in the incidence of LDH and a reduction in the average age of onset [2].

Finally, it has been shown that people with back pain have worse physical and mental health than the healthy population. It has also been shown that there is a close relationship between pathology and psychosocial stress [3].

The approach to this pathology is multidisciplinary and includes medical therapy (NSAIDs, Glucocorticoids, Opioids, Muscle Relaxants, and Antiepileptics), interventional techniques such as intraforaminal injection of corticosteroids, physiotherapy in all its forms, and finally, surgery only for selected patients. The prolonged use of drugs or infiltrations is not recommended due to the numerous side effects or, in the second case, adverse events [4].

The updated 2017 LBP guidelines from the American College of Physicians recommend the use of nonpharmacologic treatments for patients with low back pain. Physical exercise is recommended in combination with other non-pharmacological therapies such as massage therapy and physical therapy [5].

Physiotherapy is considered the first-line treatment for patients with symptoms caused by a lumbar disc herniation, both in the acute and chronic phases. Some blind randomized controlled studies have demonstrated the effectiveness of the McKenzie method, mobilizations, and vertebral tractions, even in the acute phases [6].

Furthermore, some physical therapies such as cryotherapy, electrotherapy, laser therapy, etc., have been indicated as very sensitive initial accompanying treatments [7–9].

While in the chronic phases, the effectiveness of resistance exercises for muscle strengthening of the paravertebral and abdominal muscles has been demonstrated [10].

Some nutraceuticals are considered effective in neuroprotection and pain [1–7], although controlled studies in this regard are lacking. Alpha-lipoic acid (ALA), Acetyl-L-Carnitine (ALC), and palmitoylethanolamide (PEA) are effective in the treatment of neuropathic pain from root irradiation, as is Cholecalciferol, which has recognized analgesic properties [11–15].

This prospective observational study aimed to evaluate the effectiveness of a rehabilitation program combined with the administration of ALA, ALC, Resveratrol, and Cholecalciferol in the treatment of sciatica due to herniated discs in young patients in terms of pain resolution, postural alterations, taking painkillers, and quality of life.

2. Materials and Methods

2.1. Trial Design

We conducted a randomized controlled trial on outpatients who attended the U.O.C. outpatient clinics of Functional Recovery and Rehabilitation of the A.O.U. P. Paolo Giaccone of Palermo for lumbosciatica. The study period was between September 2022 and June 2023.

The study received approval from the local ethical committee "Palermo 1" (approval no. 08/2022) of the A.O.U.P. Paolo Giaccone of Palermo and was conducted in accordance with the declaration of Helsinki. The processing of information and data has been carried out according to the guidelines of Good Clinical Practice (GCP). Clinical trials registration number NCT06078163.

2.2. Participants

The inclusion criteria used were as follows: age 18–45 years; lower back pain with NRS scale score between 5 and 7 points; symptoms attributable to sciatica which occurred no more than 4 weeks ago; pharmacological washout of NSAIDs and/or corticosteroids for at least a week; lumbar MRI examination performed no more than 3 months ago; and written consent for participation in the study. Patients were excluded if they had altered states of consciousness; sciatic pain of non-disc origin; septic states in progress; presence of scoliosis >20° of Cobb; previous spinal surgery; or were pregnant and/or breastfeeding.

2.3. Intervention

The recruited patients were randomly divided into three groups through a system of computer-generated random numbers: the Combo group (CG), composed of patients subjected to a combination of a rehabilitation protocol of 20 sessions and daily therapy with 600 mg ALA, 1000 mg ALC, 50 mg Resveratrol, and 800 UI Cholecalciferol for 30 days; the Reha group (RG), made up of patients subjected only to a rehabilitation protocol lasting 20 sessions; and the Supplement group (SG), made up of patients who took 600 mg ALA, 1000 mg ALC, 50 mg Resveratrol, and 800 IU Cholecalciferol daily for 30 consecutive days.

All patients were asked to avoid taking NSAIDs during the study period but were given the option of 500 mg of Paracetamol in combination with 30 mg of Codeine as needed in case of excessive pain.

2.4. Outcomes

All recruited patients were evaluated three times: at the time of recruitment (T0), 30 days after the start of treatment (T1), and 60 days after the end of treatment (T2). During the initial clinical evaluation, demographic information (age, sex, BMI, education level) and clinical information (smoking habits and daily working hours) were collected. For each clinical evaluation, some rating scales were administered by the same physiatrist, such as the Numeric Rating Scale (NRS) [16], to evaluate the extent of the pain; the Oswestry Disability Questionnaire (ODQ) [17], to evaluate the degree of disability caused by low back pain; and the 36-item Short Form Health Survey (SF-36) questionnaire, to assess quality of life [18].

Patients were also asked to fill in a diary where they noted any daily intake of the combination of Paracetamol and Codeine and its frequency of intake during the day.

Finally, all patients will be subjected to stabilometric analysis using a baropodometric platform that uses the FreeMed posturography system (produced by Sensor Medica, Guidonia Montecelio, Roma, Italy). During the stabilometric examination, the length of the beam (mm) was considered, which is the size of the section drawn by the oscillation of the center of gravity (CoP) during the test; the surface of the ellipse (mm^2), which includes 90% of the CoP section; the X-mean (mm), which indicates the average position maintained on the frontal plane during lateral oscillations; and the Y-mean (mm), which is the midpoint of the center of gravity on the sagittal plane during the anteroposterior oscillations [19].

2.5. Rehabilitation Protocol

The rehabilitation protocol to which the patients in the two groups (Combo and Reha) underwent was the same. It included daily sessions, 5 days a week, with a duration of 60 min, and for a total of 4 consecutive weeks. The rehabilitation protocol provided was carried out under the supervision of an experienced physiotherapist. We proceeded with an initial cardiorespiratory training phase lasting 15 min on a cycle ergometer (produced by Chinesport SPA, Udine, Italy). In the central part of the training, we proceeded with muscle strengthening exercises for the muscles of the trunk and the upper and lower part of the body and stretching exercises of the posterior kinetic chains. Each exercise was performed with 3 sets of 15 repetitions and at an intensity of 60% of the maximum. The third phase of the protocol involved therapy with physical agents, such as transcutaneous electronervous stimulation (TENS) for 20 min by stimulation with biphasic rectangular pulses of 100 microseconds and a frequency of 110 Hz, with a maximum output amplitude of 100 mA; and low power laser for 10 min with a stable, paravertebral, lumbar method with total daily doses of 18 J.

2.6. Rating Scales

The NRS scale is a quantitative rating scale by which patients are asked to rate their pain on a defined scale, from 0 to 10 [16].

The Oswestry Low Back Pain Disability Questionnaire (ODQ) (File S1) is a self-completed questionnaire with ten items covering pain intensity, ability to care for oneself, lifting and carrying, ability to walk, ability to sit, the ability to stand, the quality of sleep, social life, sexuality, and the ability to travel. Each item has six statements describing possible situations in the patient's life. The most applicable statement is checked by the patient. Questions are scored on a scale of 0 to 5. We adapted the questionnaire by omitting an item regarding sexual function. The MCID for this scale is 10 points per second [17].

The SF-36 is a questionnaire comprising eight multiple-choice questions that can be divided into two large subgroups: the physical component of the disease and the mental component of the disease. A score is assigned to each scale; the higher the score, the better the state of mental and physical health. The score ranges from 0 (worst state of health) to 100 (best state of health). The MCID for this scale is 4.9 points [18].

2.7. Statistical Methods

The data collected were indexed in Excel. The aim was to detect a mean difference in NRS (0–10) between the two groups. A power analysis was conducted with the type I error set at 0.05 and the type II error at 0.15 (85% power). The estimated sample size was 45 patients from each group to detect the minimal clinically significant difference in NRS of 2.6 units [20]. The follow-up loss was estimated to be 20%. For this reason, the numbers of 44 patients for the Combo group, 43 patients for the Reha group, and 41 for the Supplement group were considered sufficient to demonstrate our thesis.

Through the use of the Shapiro–Wilk test, the normality of our collected data was verified. In the text and the tables, we have reported continuous variables, expressed as means and standard deviations, and categorical variables, expressed as absolute numbers and percentages.

Regarding the statistical analysis of the data, we used the *t*-test for the comparison of the means between the quantitative variables, while Mood's median test was used for the comparison of the medians between the categorical variables. Finally, to evaluate the statistically significant difference between the NRS and lesion diameter variables examined between the two groups, ANOVA was used. R statistical software (R Core Team, Vienna, Austria, 2021) was used to analyze the collected data. A priori results showing $p < 0.05$ were considered statistically significant.

3. Results

Table 1 shows the general characteristics of the 128 patients who were included in the study. Participants were mainly women (59.8%), with an average age of 37.4 ± 3.4 years and an average BMI of 25.6 ± 2.8 kg/m^2. A total of 48.3% ($n = 42$) had a primary school diploma, 32.2% ($n = 28$) had a secondary school diploma, and 19.5% ($n = 17$) had a university degree. More than half of the participants (63.2%) were smokers. The average daily working hours of the recruited sample were 9.6 ± 3.2 h. The mean perceived pain was 6.2 ± 0.6 points according to the NRS scale, with a mean score on the ODQ scale of 39.6 ± 5.1 and the SF-36 scale of 56.7 ± 7.2. There were no significant differences between the participants of the three study groups regarding the different baseline characteristics analyzed (Table 1).

Table 2 shows the changes in the variables examined in the three groups at T1. In the Combo group, we observed statistically significant improvements for perceived pain (6.4 ± 0.5 vs. 3.6 ± 0.3; $p < 0.05$) and for disability (40.2 ± 4.3 vs. 33.1 ± 3.7; $p < 0.05$). In the Reha group, statistically significant improvements were observed only for disability (38.8 ± 5.2 vs. 36.3 ± 4.2; $p < 0.05$). In the Supplement group, only pain showed statistically significant improvements after 30 days of treatment for pain reduction (6.4 ± 0.6 vs. 5.5 ± 0.5; $p < 0.05$).

Table 3 shows the changes in the variables examined in the three groups at T2. The Combo group showed statistically significant improvements for pain (6.4 ± 0.5 vs. 3.2 ± 0.4; $p < 0.05$), disability (40.2 ± 4.3 vs. 34.4 ± 4.2; $p < 0.05$), and quality of life (56.4 ± 5.8 vs. 81.6 ± 6.2; $p < 0.05$). The Reha group showed statistically significant improvements for pain (6.2 ± 0.7 vs. 4.1 ± 0.6; $p < 0.05$) and disability (38.8 ± 5.2 vs. 36.1 ± 3.9; $p < 0.05$). A statistically significant improvement was also observed in the Supplement group for pain (6.4 ± 0.6 vs. 4.8 ± 0.3; $p < 0.05$) and for disability (43.5 ± 3.2 vs. 37.2 ± 4.9; $p < 0.05$).

Table 1. General patient characteristics.

Characteristics	Total ($n = 128$)	Combo Group ($n = 44$)	Reha Group ($n = 43$)	Supplement Group ($n = 41$)	p-Value
Age, mean ± SD	37.4 ± 3.4	37.7 ± 2.8	37.2 ± 4.1	37.3 ± 3.1	0.69
Sex, no. (%)					
Male	50 (39.1)	16 (36.4)	19 (44.2)	15 (36.5)	0.71
Female	78 (60.9)	28 (63.6)	24 (55.8)	26 (63.5)	
BMI, mean ± SD	25.6 ± 2.8	25.8 ± 2.1	26.3 ± 2.2	25.6 ± 3.2	0.13
Education, no. (%)					
Primary school	60 (46.8)	22 (50)	20 (46.5)	18 (43.9)	
Secondary school	38 (29.7)	15 (34.1)	13 (30.2)	10 (24.4)	0.56
Degree	30 (23.5)	7 (15.9)	10 (23.3)	13 (31.7)	
Smoker, no. (%)					
Yes	84 (65.6)	26 (59.1)	29 (67.4)	29 (70.7)	0.43
No	44 (34.4)	18 (40.9)	14 (32.6)	12 (29.3)	
Work hours, mean ± SD	9.6 ± 3.2	9.9 ± 3.1	9.5 ± 2.9	9.7 ± 2.8	0.53
NRS, mean ± SD	6.2 ± 0.6	6.4 ± 0.5	6.2 ± 0.7	6.3 ± 0.6	0.13
ODQ, mean ± SD	39.6 ± 5.1	40.2 ± 4.3	38.8 ± 5.2	39.5 ± 4.9	0.7
SF-36, mean ± SD	56.7 ± 7.2	56.4 ± 5.8	57.8 ± 6.9	55.8 ± 7.7	0.31

Table 2. Effects of the different treatments in the Combo group, in the Reha group, and in the supplement group at T1.

Characteristics	Combo Group	Reha Group	Supplement Group
NRS, mean ± SD			
T0	6.4 ± 0.5	6.2 ± 0.7	6.3 ± 0.6
T1	3.6 ± 0.3	5.3 ± 0.5	5.5 ± 0.5
p-value	<0.05	0.13	<0.05
ODQ, mean ± SD			
T0	40.2 ± 4.3	38.8 ± 5.2	39.5 ± 4.9
T1	33.1 ± 3.7	36.3 ± 4.2	42.3 ± 4.2
p-value	<0.05	<0.05	0.81
SF-36, mean ± SD			
T0	56.4 ± 5.8	57.8 ± 6.9	55.8 ± 7.7
T1	64.6 ± 4.3	60.2 ± 7.2	58.2 ± 6.5
p-value	0.19	0.12	0.48

Table 3. Effects of the different treatments in the Combo group, in the Reha group, and in the supplement group at T2.

Characteristics	Combo Group	Reha Group	Supplement Group
NRS, mean ± SD			
T0	6.4 ± 0.5	6.2 ± 0.7	6.3 ± 0.6
T1	3.2 ± 0.4	4.1 ± 0.6	4.8 ± 0.3
p-value	<0.05	<0.05	<0.05
ODQ, mean ± SD			
T0	40.2 ± 4.3	38.8 ± 5.2	39.5 ± 4.9
T1	36.1 ± 3.9	36.1 ± 3.9	37.2 ± 4.9
p-value	<0.05	<0.05	<0.05
SF-36, mean ± SD			
T0	56.4 ± 5.8	57.8 ± 6.9	55.8 ± 7.7
T1	81.6 ± 6.2	61.5 ± 8.2	58.5 ± 7.2
p-value	<0.05	0.21	0.39

Table 4 shows the comparison between the results obtained in the three groups. At T1, the Combo group showed statistically superior results compared to the other groups regarding pain ($p < 0.05$), disability ($p < 0.05$), and quality of life ($p < 0.05$). At T2, the Combo group showed statistically superior results compared to the other groups only in terms of pain ($p < 0.05$) and quality of life ($p < 0.05$). No statistically significant differences were present between the three groups at T2 for disability.

Table 4. Comparison of results at T1 and T2 between the three groups.

Characteristics	T1				T2			
	TG	CG	SG	*p*-Value	TG	CG	SG	*p*-Value
NRS, mean ± SD	3.6 ± 0.3	5.3 ± 0.5	5.5 ± 0.5	<0.05	3.2 ± 0.4	4.1 ± 0.6	4.8 ± 0.3	<0.05
ODQ, mean ± SD	33.1 ± 3.7	36.3 ± 4.2	42.3 ± 4.2	<0.05	34.4 ± 4.2	36.1 ± 3.9	37.2 ± 4.9	0.18
SF-36, mean ± SD	78.6 ± 4.3	60.2 ± 7.2	58.2 ± 6.5	<0.05	81.6 ± 6.2	61.5 ± 8.2	58.5 ± 7.2	<0.05

Table 5 shows the results of the stabilometric examination in the three groups at T1 and T2. From the analysis of these data, we only observed a statistically significant improvement at T2 in the Combo group for the average X ($p < 0.05$) compared to the other groups.

Table 5. Results of the stabilometric evaluation in the three groups.

Characteristics	T1				T2			
	CG	RG	SG	*p*-Value	CG	RG	SG	*p*-Value
Sphere length, mean ± SD	372.1 ± 124.8	374.3 ± 118.1	373.7 ± 114.5	0.93	366.2 ± 102.4	368.4 ± 105.2	368.6 ± 107.2	0.92
Ellipse surface, mean ± SD	131.1 ± 99.6	128.4 ± 98.7	129.5 ± 89.5	0.79	125.5 ± 92.5	124.6 ± 96.4	125.6 ± 94.5	0.96
Maximum oscillation, mean ± SD	2.5 ± 1.6	2.3 ± 1.5	2.3 ± 1.3	0.55	2.3 ± 1.1	2.2 ± 1.6	2.3 ± 1.4	0.73
X-media, mean ± SD	0.3 ± 7.1	0.5 ± 5.9	0.4 ± 6.3	0.98	0.2 ± 2.4	0.5 ± 5.2	0.4 ± 4.6	<0.05
Y-media, mean ± SD	−18.2 ± 11.9	−16.5 ± 10.6	−17.1 ± 10.2	0.48	−17.8 ± 10.6	−16.4 ± 10.4	−17.2 ± 9.8	0.53

Regarding the number of days of Paracetamol and Codeine intake, we observed a statistically significant reduction at T2 compared to T1 only in the treatment group (3.4 ± 0.8 vs. 1.8 ± 0.6; $p < 0.05$). At T2, however, no statistically significant difference was observed between the three groups.

4. Discussion

During this study, we tried to evaluate the effectiveness of the rehabilitation treatment combined with the administration of ALA, ALC, Resveratrol, and Cholecalciferol in the treatment of sciatica due to herniated discs in young patients. Efficacy was evaluated in terms of pain resolution, postural alterations, intake of painkillers, and quality of life.

From the results obtained, we observed how the rehabilitation treatment combined with the administration of supplements (ALA, ALC, Resveratrol, Cholecalciferol) showed statistically superior improvements in terms of reduction in pain and disability related to back pain, in the short term and above all in the long term, in the latter case also observing a significant improvement in terms of quality of life.

There are many non-pharmacological therapies for sciatic pain, and these include acupuncture, physical therapy, massage therapy, yoga, cognitive behavioral therapy or progressive relaxation, spinal manipulation, and intensive interdisciplinary rehabilitation. These treatments aim to control pain and above all to functionally recover patients, and although the level of evidence supporting the different therapies is discreet, at the moment, there is no consensus on their indication as first-choice treatments [21,22].

In addition to this, agents with antioxidant action, such as ALA and Resveratrol, have recently been identified as first-line treatment for chronic neuropathic pain [23–25] thanks to the proven efficacy compared to placebo in the treatment of neuropathic pain [26,27]. Oxidative stress that develops after peripheral neuropathic injury is considered a relevant factor responsible for neuropathic pain. It activates an inflammatory pathway that involves the entire peripheral nerve up to the spinal dorsal horn, causing sensitization and chronic neuropathic pain in the spinal column [28].

ALA can neutralize free radicals, but it also increases glutathione synthesis, regenerates other important antioxidants, and prevents the formation of glycosylated end products and mitochondrial damage from oxidative stress. Several studies have reported that ALA prevents oxidative damage to nerve tissue and nerve degeneration. Furthermore, ALA counteracts proinflammatory factors (including IL-6 and TNFα), thus reducing the overall inflammatory load. Indeed, recent reviews provide convincing evidence of the usefulness of ALA as an anti-inflammatory ally in several conditions such as acute and chronic pain, neuropathy, and ulcerative colitis [10].

Resveratrol is a widely used inhibitor of the WNT/β-catenin pathway, which plays a crucial role in many biological processes. It has shown anti-inflammatory effects in a rat arthritis model and also has well-described antioxidant properties [14].

ALC has important physiological and pharmacological actions due to its wide distribution in many tissues, including the brain [29]. Evidence from randomized controlled trials suggests that ALC is an effective agent for pain management in patients with peripheral neuropathies. The protective effects of ALC against nerve damage and pain associated with peripheral neuropathies include changes in the sensitivity of nerve growth factor (NGF) receptors, activation of M1 cholinergic muscarinic receptors in the CNS, and upregulation of glutamate receptors metabotropic type 2 (mGlu2) in dorsal root ganglion (DRG) neurons. More recently, activation of the phospholipase C (PLC)/inositol-1,4,4-triphosphate (PLC-IP 3) pathway and modulation of the transcriptional activity of nuclear factor (NF)-κB transcription factors through acetylation of the p65 subunit have been added to the list of potential mechanisms mediating the analgesic activity of ALC [13].

In line with our results, numerous studies have demonstrated how combinations of antioxidant (ALA and Resveratrol) and neurotrophic (LAC) agents can contribute to pain control, reducing the use of analgesic drugs and improving the safety profile of the treatments used [30–33].

An important observation from our research was the significant short-term reduction in the intake of painkillers. This is an important aspect to underline considering the multiple side effects resulting from their use. Among these, the most common are constipation, nausea, sedation, vomiting, and dizziness. Furthermore, with prolonged use of these drugs, many patients develop a physical and psychological dependence on opioids and the sudden cessation of the drug causes an unpleasant withdrawal syndrome (with agitation, insomnia, diarrhea, rhinorrhea, piloerection, and hyperalgesia) [34].

In the second part of our study, through stabilometric evaluation, we investigated any postural alterations that may be found in patients with sciatica. Recent studies indicate that patients with chronic low back pain have a decrease in postural control, manifesting balance problems. Postural balance is controlled by sensory information, central processing, and neuromuscular responses. Alterations in proprioception are identified as one of the possible causes of alterations in postural balance in subjects with low back pain. This type of pain is associated with decreased proprioception and muscle strength, which can affect the quality of information and compromise the relationship between postural responses and sensory responses to information [35].

Our data showed that the group subjected to the rehabilitation treatment combined with the administration of ALA, Resveratrol, LAC, and Cholecalciferol presented fewer perturbations in the frontal plane in the long term, without showing particularly significant improvements in all the other stabilometric parameters, both short and long term.

Pain represents the main factor responsible for changes in postural control, regardless of the intensity of the pain. This determines an alteration of the upright position which determines an increased activation of the lumbar muscles, with consequent increase in muscle fatigue [35–37].

In addition to pain, the impaired postural control found in patients with sciatic pain could result from a reduction in proprioceptive acuity, restrictive trunk movement, and protective trunk muscle strategies [38]. With the development of chronicity, a progressive decrease in variability and an increase in rigidity would lead to an increase in postural sway [38].

Several systematic reviews [35,38] suggest that low back pain leads to a shift in postural control from the lumbar spine to the ankles, and postural control at the ankles increases the magnitude of swing. However, other studies have reported smaller and slower CoP movements during quiet standing with eyes closed and eyes open [39].

Lemos et al. [40] analyzed the influence of low back pain on the balance of athletes of the Brazilian women's canoe team and found an increase in the extent of displacement of the CoP in the horizontal plane in athletes with the presence of pain. Park et al. observed that, compared to controls, individuals with cLBP showed increased postural sway during quiet standing. In contrast, we did not find strong support for an effect of cLBP on postural swing velocity.

Our study was not without limitations. The first is represented by the small sample size which does not allow the results obtained to be generalized. Another limitation may be represented by the lack of evaluation between the different entities of lumbar pain and the postural alterations present in patients with sciatica. Finally, a further limitation may be represented by the failure to evaluate the stabilometric examination with eyes closed.

5. Conclusions

For the treatment of sciatica, the combined treatment with rehabilitation and supplements with anti-inflammatory, pain-relieving, and antioxidant action is effective in reducing pain and improving disability in the short term, compared to a single rehabilitation treatment or with supplements. In the long term, the combination also allows an improvement in the quality of life. Finally, the intake of natural substances reduces the need for painkillers.

Supplementary Materials: The following supporting information can be downloaded at: https://www.mdpi.com/article/10.3390/medicina59122197/s1, File S1: The Oswestry Low Back Pain Disability Questionnaire (ODQ).

Author Contributions: Conceptualization, F.V. and D.S.; methodology, F.V.; software, S.T.; validation, D.S., S.R. and G.L.M.; formal analysis, C.S.; investigation, F.V.; resources, F.V.; data curation, D.S.; writing—original draft preparation, F.V.; visualization, D.S.; supervision, S.R. and G.L.M. All authors have read and agreed to the published version of the manuscript.

Funding: This research received no external funding.

Institutional Review Board Statement: The study was conducted in accordance with the Declaration of Helsinki and approved by the Ethics Committee of Palermo 1 (Protocol Code 08/2022 and date of approval 19 August 2022).

Informed Consent Statement: Written informed consent was obtained from the patients to publish this paper.

Data Availability Statement: The data used to support the findings of this study are available from the corresponding author upon request.

Conflicts of Interest: The authors declare no conflict of interest.

References

1. Urits, I.; Burshtein, A.; Sharma, M.; Testa, L.; Gold, P.A.; Orhurhu, V.; Viswanath, O.; Jones, M.R.; Sidransky, M.A.; Spektor, B.; et al. Low Back Pain, a Comprehensive Review: Pathophysiology, Diagnosis, and Treatment. *Curr. Pain Headache Rep.* **2019**, *23*, 23. [CrossRef]

2. Wong, T.; Patel, A.; Golub, D.; Kirnaz, S.; Goldberg, J.L.; Sommer, F.; Schmidt, F.A.; Nangunoori, R.; Hussain, I.; Härtl, R. Prevalence of Long-Term Low Back Pain after Symptomatic Lumbar Disc Herniation. *World Neurosurg.* **2023**, *170*, 163–173.e1. [CrossRef] [PubMed]

3. Yang, H.; Lu, M.; Haldeman, S.; Swanson, N. Psychosocial risk factors for low back pain in US workers: Data from the 2002–2018 quality of work life survey. *Am. J. Ind. Med.* **2023**, *66*, 41–53. [CrossRef] [PubMed]

4. Oliveira, C.B.; Oliveira, C.B.; Maher, C.G.; Maher, C.G.; Pinto, R.Z.; Pinto, R.Z.; Traeger, A.C.; Traeger, A.C.; Lin, C.-W.C.; Lin, C.-W.C.; et al. Clinical practice guidelines for the management of non-specific low back pain in primary care: An updated overview. *Eur. Spine J.* **2018**, *27*, 2791–2803. [CrossRef] [PubMed]

5. Qaseem, A.; Wilt, T.J.; McLean, R.M.; Forciea, M.A. Clinical Guidelines Committee of the American College of Physicians Noninvasive Treatments for Acute, Subacute, and Chronic Low Back Pain: A Clinical Practice Guideline from the American College of Physicians. *Ann. Intern. Med.* **2017**, *166*, 514–530. [CrossRef] [PubMed]

6. Maher, C.; Underwood, M.; Buchbinder, R. Non-specific low back pain. *Lancet* **2017**, *389*, 736–747. [CrossRef] [PubMed]

7. Salas-Fraire, O.; Rivera-Pérez, J.A.; Guevara-Neri, N.P.; Urrutia-García, K.; Martínez-Gutiérrez, O.A.; Salas-Longoria, K.; Morales-Avalos, R. Efficacy of whole-body cryotherapy in the treatment of chronic low back pain: Quasi-experimental study. *J. Orthop. Sci.* **2023**, *28*, 112–116. [CrossRef] [PubMed]

8. Johnson, M.I.; Paley, C.A.; Jones, G.; Mulvey, M.R.; Wittkopf, P.G. Efficacy and safety of transcutaneous electrical nerve stimulation (TENS) for acute and chronic pain in adults: A systematic review and meta-analysis of 381 studies (the meta-TENS study). *BMJ Open* **2022**, *12*, e051073. [CrossRef]

9. Glazov, G.; Yelland, M.; Emery, J. Low-Level Laser Therapy for Chronic Non-Specific Low Back Pain: A Meta-Analysis of Randomised Controlled Trials. *Acupunct. Med.* **2016**, *34*, 328–341. [CrossRef]

10. Wang, W.; Long, F.; Wu, X.; Li, S.; Lin, J. Clinical Efficacy of Mechanical Traction as Physical Therapy for Lumbar Disc Herniation: A Meta-Analysis. *Comput. Math. Methods Med.* **2022**, *2022*, 5670303. [CrossRef]

11. Scaturro, D.; Asaro, C.; Lauricella, L.; Tomasello, S.; Varrassi, G.; Mauro, G.L. Combination of Rehabilitative Therapy with Ultramicronized Palmitoylethanolamide for Chronic Low Back Pain: An Observational Study. *Pain Ther.* **2020**, *9*, 319–326. [CrossRef] [PubMed]

12. Costantino, M.; Guaraldi, C.; Costantino, D.; De Grazia, S.; Unfer, V. Peripheral neuropathy in obstetrics: Efficacy and safety of α-lipoic acid supplementation. *Eur. Rev. Med. Pharmacol. Sci.* **2014**, *18*, 2766–2771. [PubMed]

13. Sarzi-Puttini, P.; Giorgi, V.; Di Lascio, S.; Fornasari, D. Acetyl-L-carnitine in chronic pain: A narrative review. *Pharmacol. Res.* **2021**, *173*, 105874. [CrossRef] [PubMed]

14. Genovese, T.; Impellizzeri, D.; D'amico, R.; Cordaro, M.; Peritore, A.F.; Crupi, R.; Gugliandolo, E.; Cuzzocrea, S.; Fusco, R.; Siracusa, R.; et al. Resveratrol Inhibition of the WNT/β-Catenin Pathway following Discogenic Low Back Pain. *Int. J. Mol. Sci.* **2022**, *23*, 4092. [CrossRef] [PubMed]

15. Zadro, J.; Shirley, D.; Ferreira, M.; Carvalho-Silva, A.P.; Lamb, E.S.; Cooper, C.; Ferreira, P.H. Mapping the Association between Vitamin D and Low Back Pain: A Systematic Review and Meta-Analysis of Observational Studies. *Pain Physician* **2017**, *20*, 611–640. [CrossRef] [PubMed]

16. Lewis, C.; Stjernbrandt, A.; Wahlström, J. The association between cold exposure and musculoskeletal disorders: A prospective population-based study. *Int. Arch. Occup. Environ. Health* **2023**, *96*, 565–575. [CrossRef]

17. Smeets, R.; Köke, A.; Lin, C.; Ferreira, M.; Demoulin, C. Measures of function in low back pain/disorders: Low Back Pain Rating Scale (LBPRS), Oswestry Disability Index (ODI), Progressive Isoinertial Lifting Evaluation (PILE), Quebec Back Pain Disability Scale (QBPDS), and Roland-Morris Disability Questionnaire (RDQ). *Arthritis Care Res.* **2011**, *63* (Suppl. S11), S158–S173. [CrossRef]

18. Kerr, D.; Zhao, W.; Lurie, J.D. What Are Long-term Predictors of Outcomes for Lumbar Disc Herniation? A Randomized and Observational Study. *Clin. Orthop. Relat. Res.* **2015**, *473*, 1920–1930. [CrossRef]

19. Scaturro, D.; Rizzo, S.; Sanfilippo, V.; Giustino, V.; Messina, G.; Martines, F.; Falco, V.; Cuntrera, D.; Moretti, A.; Iolascon, G.; et al. Effectiveness of Rehabilitative Intervention on Pain, Postural Balance, and Quality of Life in Women with Multiple Vertebral Fragility Fractures: A Prospective Cohort Study. *J. Funct. Morphol. Kinesiol.* **2021**, *6*, 24. [CrossRef]

20. Ogura, Y.; Ogura, K.; Kobayashi, Y.; Kitagawa, T.; Yonezawa, Y.; Takahashi, Y.; Yoshida, K.; Yasuda, A.; Shinozaki, Y.; Ogawa, J. Minimum clinically important difference of major patient-reported outcome measures in patients undergoing decompression surgery for lumbar spinal stenosis. *Clin. Neurol. Neurosurg.* **2020**, *196*, 105966. [CrossRef]

21. Chou, R.; Qaseem, A.; Snow, V.; Casey, D.; Cross, J.T.; Shekelle, P.; Owens, D.K. Clinical Efficacy Assessment Subcommittee of the American College of Physicians and the American College of Physicians/American Pain Society Low Back Pain Guidelines Panel* Diagnosis and Treatment of Low Back Pain: A Joint Clinical Practice Guideline from the American College of Physicians and the American Pain Society. *Ann. Intern. Med.* **2007**, *147*, 478–491, Erratum in *Ann. Intern. Med.* **2008**, *148*, 247–248. [CrossRef]

22. National Institute for Health and Care Excellence. *Neuropathic Pain in Adults: Pharmacological Management in Non-Specialist Settings*; National Institute for Health and Care Excellence (NICE): London, UK, 2020.

23. Watson, J.C.; Dyck, P.J.B. Peripheral Neuropathy: A Practical Approach to Diagnosis and Symptom Management. *Mayo Clin. Proc.* **2015**, *90*, 940–951. [CrossRef] [PubMed]

24. Tao, L.; Ding, Q.; Gao, C.; Sun, X. Resveratrol attenuates neuropathic pain through balancing pro-inflammatory and anti-inflammatory cytokines release in mice. *Int. Immunopharmacol.* **2016**, *34*, 165–172. [CrossRef] [PubMed]

25. Xie, J.; Liu, S.; Wu, B.; Li, G.; Rao, S.; Zou, L.; Yi, Z.; Zhang, C.; Jia, T.; Zhao, S.; et al. The protective effect of resveratrol in the transmission of neuropathic pain mediated by the P2X7 receptor in the dorsal root ganglia. *Neurochem. Int.* **2017**, *103*, 24–35. [CrossRef] [PubMed]
26. Ranieri, M.; Sciuscio, M.; Cortese, A.; Santamato, A.; Di Teo, L.; Ianieri, G.; Bellomo, R.; Stasi, M.; Megna, M. The Use and Alpha-Lipoic Acid (ALA), Gamma Linolenic Acid (GLA) and Rehabilitation in the Treatment of Back Pain: Effect on Health-Related Quality of Life. *Int. J. Immunopathol. Pharmacol.* **2009**, *22* (Suppl. S3), 45–50. [CrossRef] [PubMed]
27. Memeo, A.; Loiero, M. Thioctic Acid and Acetyl-L-Carnitine in the Treatment of Sciatic Pain Caused by a Herniated Disc: A randomized, double-blind, comparative study. *Clin. Drug Investig.* **2008**, *28*, 495–500. [CrossRef]
28. Checchia, G.A.; Mauro, G.L.; Morico, G.; Oriente, A.; Lisi, C.; Polimeni, V.; Lucia, M.; Ranieri, M. Management of Peripheral Neuropathies Study Group Observational multicentric study on chronic sciatic pain: Clinical data from 44 Italian centers. *Eur. Rev. Med. Pharmacol. Sci.* **2017**, *21*, 1653–1664.
29. Scaturro, D.; Vitagliani, F.; Di Bella, V.E.; Falco, V.; Tomasello, S.; Lauricella, L.; Mauro, G.L. The Role of Acetyl-Carnitine and Rehabilitation in the Management of Patients with Post-COVID Syndrome: Case-Control Study. *Appl. Sci.* **2022**, *12*, 4084. [CrossRef]
30. Nakajima, S.; Ohsawa, I.; Nagata, K.; Ohta, S.; Ohno, M.; Ijichi, T.; Mikami, T. Oral supplementation with melon superoxide dismutase extract promotes antioxidant defences in the brain and prevents stress-induced impairment of spatial memory. *Behav. Brain Res.* **2009**, *200*, 15–21. [CrossRef]
31. Yasui, K.; Baba, A. Therapeutic potential of superoxide dismutase (SOD) for resolution of inflammation. *Inflamm. Res.* **2006**, *55*, 359–363. [CrossRef]
32. Bertolotto, F.; Massone, A. Combination of Alpha Lipoic Acid and Superoxide Dismutase Leads to Physiological and Symptomatic Improvements in Diabetic Neuropathy. *Drugs R&D* **2012**, *12*, 29–34. [CrossRef]
33. Mauro, G.L.; Cataldo, P.; Barbera, G.; Sanfilippo, A. α-Lipoic Acid and Superoxide Dismutase in the Management of Chronic Neck Pain: A Prospective Randomized Study. *Drugs R&D* **2014**, *14*, 1–7. [CrossRef]
34. Deyo, R.A.; Von Korff, M.; Duhrkoop, D. Opioids for low back pain. *BMJ* **2015**, *350*, g6380. [CrossRef] [PubMed]
35. Park, J.; Nguyen, V.Q.; Ho, R.L.M.; Coombes, S.A. The effect of chronic low back pain on postural control during quiet standing: A meta-analysis. *Sci. Rep.* **2023**, *13*, 7928. [CrossRef] [PubMed]
36. Vogt, L.; Pfeifer, K.; Banzer, W. Neuromuscular control of walking with chronic low-back pain. *Man. Ther.* **2003**, *8*, 21–28. [CrossRef] [PubMed]
37. Brumagne, S.; Janssens, L.; Knapen, S.; Claeys, K.; Suuden-Johanson, E. Persons with recurrent low back pain exhibit a rigid postural control strategy. *Eur. Spine J.* **2008**, *17*, 1177–1184. [CrossRef] [PubMed]
38. Braga, A.B.; Rodrigues, A.C.d.M.A.; de Lima, G.V.M.P.; de Melo, L.R.; de Carvalho, A.R.; Bertolini, G.R.F. Comparison of static postural balance between healthy subjects and those with low back pain. *Acta Ortop. Bras.* **2012**, *20*, 210–212. [CrossRef]
39. Brumagne, S.; Cordo, P.; Verschueren, S. Proprioceptive weighting changes in persons with low back pain and elderly persons during upright standing. *Neurosci. Lett.* **2004**, *366*, 63–66. [CrossRef]
40. Lemos, L.F.C.; Teixeira, C.S.; Mota, C.B. Lombalgia e o equilíbrio corporal de atletas da seleção brasileira feminina de canoagem velocidade. *Rev. Bras. Cineantropom Desempenho.* **2010**, *12*, 457–463. [CrossRef]

Article

Characteristics of Age Classification into Five-Year Intervals to Explain Sarcopenia and Immune Cells in Older Adults

Seung-Jae Heo [1],* and Yong-Seok Jee [2],*

[1] Department of Life Sports Education, Kongju National University, Gongju-si 32588, Republic of Korea
[2] Research Institute of Sports and Industry Science, Hanseo University, Seosan-si 31962, Republic of Korea
*   Correspondence: hak226@kongju.ac.kr (S.-J.H.); jeeys@hanseo.ac.kr (Y.-S.J.); Tel.: +82-10-8722-1401 (S.-J.H.); +82-41-660-1028 (Y.-S.J.)

Abstract: *Background and Objectives*: This study focused on investigating sarcopenic factors and immune cells in older adulthood. To achieve this, the variables related to sarcopenia and immune cells in people living in the same community were analyzed. *Materials and Methods*: A total of 433 elderly individuals aged 61 to 85 years were randomly categorized as follows in 5-year intervals: 68 in the youngest-old group (aged 61–65), 168 in the young-old group (aged 66–70), 127 in the middle-old group (aged 71–75), 46 in the old-old group (aged 76–80), and 19 in the oldest-old group (aged 81–85). *Results*: With the progression of age, calf circumference (-8.4 to -11.05%; $p = 0.001$) and grip strength (-9.32 to -21.01%; $p = 0.001$) exhibited a noticeable reduction with each successive 5-year age bracket. Conversely, the capability to complete the five-time chair stand demonstrated a clear incline (32.49 to 56.81%; $p = 0.001$), starting from the middle-aged group. As for appendicular skeletal muscle mass, there was an evident tendency for it to decrease (-7.08 to -26.62%; $p = 0.001$) with increasing age. A gradual decline in natural killer cells became apparent within the old-old and oldest-old groups (-9.28 to -26.27%; $p = 0.001$). The results of the post hoc test revealed that CD3 T cells showcased their peak levels in both the youngest-old and young-old groups. This was followed by the middle-old and old-old groups, with slightly lower levels. This pattern was similarly observed in CD4 T cells, CD8 T cells, and CD19 B cells. *Conclusions*: This study reaffirmed that sarcopenia and immune cell function decline with each successive 5-year increase in age. Considering these findings, the importance of implementing programs aimed at ensuring a high-quality extension of life for the elderly is strongly underscored.

Keywords: sarcopenia; calf circumference; grip strength; appendicular skeletal muscle mass; natural killer cell; immune cells

Citation: Heo, S.-J.; Jee, Y.-S. Characteristics of Age Classification into Five-Year Intervals to Explain Sarcopenia and Immune Cells in Older Adults. *Medicina* **2023**, *59*, 1700. https://doi.org/10.3390/medicina 59101700

Academic Editors: Milica Lazovic and Dejan Nikolić

Received: 21 August 2023
Revised: 19 September 2023
Accepted: 21 September 2023
Published: 22 September 2023

1. Introduction

By the year 2030, it is estimated that the number of people aged 60 years and older will reach 1.4 billion, and by 2050, it is expected to further grow to 2.1 billion [1]. Because of this aging population, there is a growing need for policies and initiatives that cater to the specific needs and challenges faced by older adults to ensure their well-being and quality of life [2]; this is viewed as a significant step towards prioritizing efforts to promote the well-being and health of older adults worldwide [3].

Adjusting age group categorization is one way to ensure that healthcare and research efforts align with the specific needs and characteristics of older individuals within distinct age ranges [4]. Given the wide range of health conditions and the rising prevalence of chronic diseases, it can be recommended that most research employing 5-year age intervals may yield beneficial outcomes [5]. By using 5-year age intervals, researchers and healthcare practitioners can better account for the emergence of various chronic conditions and the potential differences in health status within specific age ranges. For instance, considering the survival rate of cancer survivors is essential, as it can impact the health status and needs of older adults within specific age brackets to improve the health condition of older adults [6].

Sarcopenia, which involves the loss of muscle mass and strength, can significantly reduce quality of life if left untreated or poorly managed. Sarcopenia can lead to mobility issues and increased dependency, which can adversely affect an individual's physical fitness and performance. Addressing sarcopenia and other age-related health conditions becomes crucial to mitigate their negative impact on quality of life [7]. Understanding the factors contributing to sarcopenia's progression can pave the way for the development of more effective preventive and treatment strategies. By taking proactive measures and intervening early, healthcare providers can potentially reduce the burden of sarcopenia and enhance the quality of life of older adults, promoting healthier and more fulfilling lives as people age. Sarcopenia can be categorized as a geriatric syndrome, which is defined by the Asian Working Group for Sarcopenia (AWGS) as an "age-related loss of skeletal muscle mass plus loss of muscle strength and/or reduced physical performance" [7,8]. It is also considered a generalized and progressive skeletal muscle disorder that is associated with an increased likelihood of adverse outcomes by the European Working Group on Sarcopenia in Older People 2 (EWGSOP2) [9]. In particular, the decrease in muscle mass that occurs with sarcopenia is accompanied by a decline in immune cell function, leading to an increased susceptibility to various infections [10].

The immune system has been suggested to play a role in affecting myogenesis, as indicated in many studies [11,12]. While significant progress has been made in understanding the interaction between the skeletal muscle and the immune system, the impact of health status within specific age ranges on this relationship remains relatively unexplored. Further research in this area could provide valuable insights into the underlying mechanisms of sarcopenia and the immune system in aged adults. Consequently, it is essential to investigate variations in sarcopenia and immunocytes based on various age categorizations. Such detailed age-specific categorization can make it easier to prevent muscle atrophy in the elderly and can be considered a proactive measure to prevent the decline in immune cell function that comes with aging. This can contribute to maintaining a high quality of life and extending the lifespan. Therefore, the primary objective of this study was to examine and compare the characteristics of sarcopenia and immunocytes by categorizing the elderly into age groups with intervals of 5 years.

2. Materials and Methods

2.1. Study Design and Participants

This research took place in June and July 2023. This single-blind, randomized, prospective cohort study was conducted in a research center at Seoul Seniors Towers, Korea. Through promotional activities at the Seoul Senior Tower, individuals who voluntarily enlisted in the study were expected to be between the ages of 60 and 100 and capable of independent living. Participants were unaware of the randomization results. The variables to be investigated in this study were determined at the time of analysis. The randomization list was generated by a computer using block randomization with five blocks. Prior to the study, the principal investigator explained all the procedures to the participants in detail. All participants read and signed an informed consent form. This study followed the principles of the Declaration of Helsinki and received approval from the institutional ethics committee (HS23-08-02, 2 November 2022 to 1 November 2023). This study was also registered with the Clinical Research Information Service, reference KCT0008712. This study determined the required sample size for the ANOVA (fixed effects, omnibus, one-way) using G*Power software (ver. 3.1.9.7; Heinrich-Heine-Universität Düsseldorf, Düsseldorf, Germany). The levels of sarcopenia factors and immune cells were then examined. The inclusion criteria comprised individuals without cancer or severe illnesses, those capable of performing a certain level of physical activity, and individuals who had not undergone surgery within the past 3 months. Older adults with severe musculoskeletal disorders, lung disease, cancer, and cachexia, which could potentially influence muscle mass and function, were excluded from the study [13].

After eliminating 139 individuals from the initial pool of 572 eligible participants, the remaining 433 participants (males, 53.5%; females, 46.5%) were distributed across five distinct groups. Among the 69 individuals belonging to the youngest-old category (aged 61–65 years), 1 person was unavailable during the follow-up stage. Within the young-old group (aged 66–70 years), consisting of 169 individuals, 1 participant was also lost during the follow-up phase. In the middle-old group, which consisted of 128 individuals aged 71–75 years, 1 individual was also not included in the follow-up phase. Notably, there were no dropouts in the old-old group (aged 76–80 years), which consisted of 46 individuals. Among the 21 individuals in the oldest-old group (aged 81–85 years), 2 participants were lost during the follow-up phase, as illustrated in Figure 1.

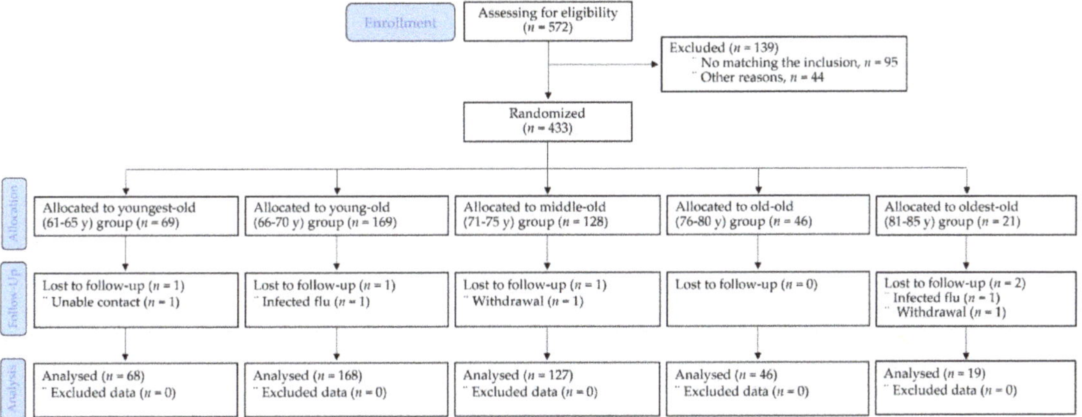

Figure 1. Participants' allocation (consolidated standards for reporting of trials flow diagram).

2.2. Daily Physical Activity Measurements

The participants' daily physical activity levels were assessed and quantified using the International Physical Activity Questionnaire (IPAQ)—shortened form version [14]. Participants filled out questionnaires based on their weekly physical activity records, and daily calorie expenditure was determined by calculating metabolic equivalent minutes.

2.3. Anthropometric Measurements

All participants underwent an assessment of their height, body weight, body mass index (BMI), lean mass, and fat mass. These measurements were conducted using a body composition analyzer known as the Inbody 770 (Biospace Co., Ltd., Seoul, Republic of Korea). This analyzer employs the bioelectrical impedance analysis (BIA) technique to offer insight into the distribution of various tissues within the body.

2.4. Sarcopenia Measures

Following the AWGS guidelines [7], the first step involved evaluating calf circumference. The participants' left knees were bent to achieve a 90-degree angle, and a measuring tape was placed around the calf area. The tape was then gently moved along the calf to identify the point of greatest circumference, with care taken to avoid compressing the subcutaneous tissue [15]. The calf circumference values were collected from both sides and averaged to establish the ultimate calf circumference measurement. The established cutoff values were <34 cm for males and <33 cm for females. Next, the assessment involved measuring grip strength, accomplished using a grip dynamometer (TKK 5401; Takei, Tokyo, Japan). Participants were instructed to stand with their feet at shoulder width and comfortably adjust the dynamometer to fit their hand's length, and then squeeze it with their maximum force. Each participant's right and left hands underwent two tests each, and the average of these values was employed for subsequent analysis. The established cutoff values for assessing sarcopenia based on grip strength were <28 kg for males and <18 kg for females, as indicated by previous studies [7,16]. In the third step, the participants' physical performance was gauged through the five-time chair stand test. In this assessment, participants were required to rise from a chair five times sequentially, with their arms folded across their chest. The time taken to complete the task was recorded in seconds [16]. The established diagnostic criterion for identifying sarcopenia using this test is a duration of ≥ 12 s for both males and females, as outlined in prior research [7]. The dual-energy X-ray absorptiometry (DXA) technique (Lunar Co., Madison, WI, USA), was employed to determine the appendicular skeletal muscle (ASM) mass and subsequently diagnose sarcopenia. In this procedure, elderly participants assumed a comfortable lying position, and a full-body scan was conducted. The scanning process generally spanned around 15 to 20 min. To compute the ASM mass using DXA, the muscle masses from both the arms and legs were aggregated and subsequently divided by the square of the participant's height in meters. The established criteria for diagnosing sarcopenia based on ASM mass are 7 kg/m^2 or below for males and 5.4 kg/m^2 or below for females [7].

2.5. Immunocyte Measures

The blood collection and analysis process in the study involved several steps and utilized specific tubes and techniques for different measurements. (1) Blood Collection: Participants underwent a fasting period of 10 h before blood samples were collected from the median cubital vein. (2) Differential Blood Cell Counts. (3) Flow Cytometric Assays: Flow cytometry was performed to assess cellular immune function. (4) Serum Sample Processing: After collection, the serum samples were allowed to clot for 30–45 min in an upright position; they were then subjected to centrifugation to separate the serum from the blood cells. (5) Quantification of Immunocytes: To evaluate cellular immune function, the quantification of immunocytes was performed using flow cytometry; specific antibodies were used to target different cell types, including CD56+ NK cells, CD3+ T-cells, CD4+ helper T-cells, CD8+ cytotoxic T-cells, and CD19+ B-cells. The use of specific tubes and techniques ensured accurate and reliable measurements of different parameters [17]. Natural killer cells have the ability to identify infected or cancerous cells, binding to them, and then releasing enzymes and other substances that effectively break down the outer membrane of these cells [18]. These cells play a significant role in the early defense against viral infections. At birth, acquired immunity is not present; instead, it develops through a learning process. This learning process begins when an individual's immune system encounters a foreign intruder and identifies its antigen. Subsequently, the adaptive immune components learn the most effective way to combat this antigen and commence building a memory for it [19]. Acquired immunity is also known as specific immunity because it orchestrates a tailored response against specific antigens encountered in the past. As we grow older, our immune system tends to lose some of its ability to differentiate between self and non-self. Consequently, autoimmune diseases may manifest more frequently. For instance, the quantity of white blood cells and their subsets capable of responding to new antigens diminishes, which, in turn, diminishes the body's capacity to retain memory and mount a defense when encountering new antigens [20].

2.6. Data Analyses

The data were analyzed using SPSS software (version 25; IBM Corp., Armonk, NY, USA). Graphical representations of the results were generated using GraphPad Prism 10.0 (La Jolla, CA, USA). Descriptive statistics, such as mean and standard deviation, were used to summarize the data. Categorical variables were presented as counts (*n*) and percentages (%) and compared using the χ^2 test. The Shapiro–Wilk test was utilized to evaluate the regularity of anthropometric and clinical measurements. The equality of variances was assessed, and the mean values were analyzed using one-way ANOVA when assuming equal variances. In cases where equal variances were not assumed, the Kruskal–Wallis test was employed. Additionally, ANCOVA verification was performed to assess the factors related to sarcopenia and immune cell variables with respect to age. To identify noteworthy protocol impacts, the Scheffe test was performed for further analysis. In addition, this study calculated delta values to assess differences between the youngest-old and young-old groups, between the young-old and middle-old groups, between the middle-old and old-old groups, and between the old-old and oldest-old groups, respectively. Furthermore, this study conducted an analysis using the Kruskal–Wallis test to examine intergroup differences in these delta values. Effect sizes were determined by converting partial eta-squared to Cohen's d [21]; values were classified as small ($0.00 \leq d \leq 0.49$), medium ($0.50 \leq d \leq 0.79$), and large ($d \leq 0.80$). A significance level of $p \leq 0.05$ was used to determine significance.

3. Results

3.1. Demographic, Anthropometric, and Clinical Characteristics

As indicated in Table 1, there was a notable variation in the age of the participants across the groups. In the youngest-old category group, there were 44 males and 24 females. The young-old group had 89 males and 79 females. Within the middle-old group, there were 62 males and 65 females. The old-old group had 25 males and 21 females. Among the 19 individuals in the oldest-old group, there were 9 males and 10 females. The gender distribution did not display any significant divergence among the five groups. In a characteristic manner, lean mass exhibited a decline with increasing age, whereas fat mass demonstrated no distinction between the youngest-old and young-old groups. Subsequently, there was an upswing in fat mass after the middle-old group.

Table 1. Demographic, anthropometric, and clinical characteristics of the old population.

	Groups					p	η^2
Items	**Youngest-Old** ($n = 68$)	**Young-Old** ($n = 168$)	**Middle-Old** ($n = 127$)	**Old-Old** ($n = 46$)	**Oldest-Old** ($n = 19$)		
Age (y)	64.46 ± 1.11	67.57 ± 1.32	73.17 ± 1.38	77.61 ± 1.27	83.00 ± 1.53	0.001	0.933
Height (cm)	170.18 ± 4.30	169.32 ± 4.30	168.85 ± 5.00	167.85 ± 4.73	167.01 ± 4.13	0.003	0.039
Body weight (kg)	70.60 ± 6.37	73.84 ± 7.66	73.24 ± 7.96	73.55 ± 7.28	69.29 ± 8.04	0.009	0.032
BMI (kg/m^2)	24.39 ± 2.23	25.79 ± 2.85	26.07 ± 3.39	25.84 ± 2.80	24.87 ± 3.06	0.002	0.038
Lean mass (kg)	48.65 ± 6.08	47.25 ± 5.21	47.20 ± 5.85	45.65 ± 4.36	41.71 ± 4.57	0.001	0.061
Fat mass (kg)	22.98 ± 3.46	22.60 ± 2.82	24.39 ± 3.32	26.37 ± 2.82	26.21 ± 3.16	0.001	0.156
OV/OB	0.44 ± 0.50	0.49 ± 0.50	0.55 ± 0.50	0.61 ± 0.49	0.63 ± 0.50	0.285	0.012
DM	0.18 ± 0.38	0.21 ± 0.41	0.18 ± 0.39	0.13 ± 0.34	0.16 ± 0.37	0.741	0.005
HTN	0.24 ± 0.43 [c]	0.35 ± 0.48 [b]	0.53 ± 0.50 [b]	0.61 ± 0.49 [a]	0.63 ± 0.50 [a]	0.001	0.069
HLD	0.31 ± 0.47	0.30 ± 0.46	0.33 ± 0.47	0.22 ± 0.42	0.32 ± 0.48	0.717	0.005
ARTH	0.32 ± 0.47	0.51 ± 0.50	0.54 ± 0.50	0.54 ± 0.50	0.58 ± 0.51	0.041	0.023
LBP	0.26 ± 0.44 [d]	0.32 ± 0.47 [bc]	0.54 ± 0.50 [bc]	0.61 ± 0.49 [b]	0.68 ± 0.48 [a]	0.001	0.078

All data represent mean $\pm$ standard deviation. Symbols [a, b, c,] and [d] represent post hoc results from the Scheffe test. BMI, body mass index; OV, overweight; OB, obesity; DM, diabetes mellitus; HTN, hypertension; HLD, hyperlipidemia; ARTH, arthritis; LBP, low back pain.

The older adults who took part in this study commonly shared conditions such as overweight, obesity, diabetes mellitus, hypertension, hyperlipidemia, arthritis, and low back pain. Regarding clinical traits, only hypertension, arthritis, and low back pain displayed noteworthy variations between the groups. In contrast, no substantial disparities were observed between the groups in relation to obesity, diabetes, and hyperlipidemia. The outcomes of the post hoc test revealed that hypertension was least prevalent within the youngest-old group and that it displayed a gradual escalation with advancing age. This trend indicated a rise in the frequency of hypertension cases following the old-old group. With the progression of age, there is a corresponding increase in the occurrence of low back pain among elderly individuals, with the highest prevalence observed within the oldest-old group. In terms of daily calorie expenditure from physical activities, the average calories burned through these activities were as follows: 1740.35 ± 289.28 MET·min/week for the youngest-old group, 1627.90 ± 233.20 MET·min/week for the young-old group, 1343.36 ± 187.45 MET·min/week for the middle-old group, 1211.28 ± 289.28 MET·min/week for the old-old group, and 1132.58 ± 208.81 MET·min/week for the oldest-old group. Significant differences were observed among these groups ($X^2 = 181.708$; $p = 0.001$; $\eta^2 = 0.422$).

3.2. Comparative Results of Sarcopenic Factors

As indicated in Table 2, every assessed parameter associated with sarcopenia exhibited notable distinctions across all five groups. Notably, there was a discernible decline in calf circumference and grip strength with each successive 5-year increment in age. Conversely, the ability to perform the five-time chair stand displayed evident deterioration from the middle-aged group onwards. Regarding ASM, there was a noticeable tendency for it to diminish with advancing age. Importantly, a significant reduction was observed, starting from the young-old group and beyond. When analyzing calf circumference with age as a covariate, there was no significant difference observed among the groups (F = 1.743; $p = 0.140$). However, handgrip strength exhibited a significant difference among the groups (F = 7.891; $p = 0.001$). When analyzing the five-time chair stand with age as a covariate, a significant difference was observed among the groups (F = 2.763; $p = 0.027$); similarly, the ASM also displayed a significant difference among the groups (F = 36.898; $p = 0.001$).

Table 2. Age-dependent disparities in sarcopenic factors.

	Groups					p	η^2
	Youngest-Old (*n* = 68)	Young-Old (*n* = 168)	Middle-Old (*n* = 127)	Old-Old (*n* = 46)	Oldest-Old (*n* = 19)		
Calf circumference (cm)	29.44 ± 4.03	26.92 ± 3.18	24.66 ± 3.17	21.93 ± 3.46	19.90 ± 2.84	0.001	0.357
Grip strength (kg)	28.74 ± 5.97	22.89 ± 5.47	20.75 ± 5.13	16.39 ± 4.21	13.65 ± 2.05	0.001	0.348
5-time chair stand test (sec)	12.40 ± 1.63	16.43 ± 3.33	25.77 ± 6.04	30.65 ± 5.68	39.16 ± 4.30	0.001	0.723
ASM (kg/m^2)	10.69 ± 2.05	7.84 ± 1.44	6.65 ± 1.16	6.67 ± 1.32	6.20 ± 0.51	0.001	0.488

All data represent mean ± standard deviation. ASM, appendicular skeletal muscle mass.

After post hoc analysis, calf circumference was the thickest in the youngest-old group and gradually decreased significantly with increasing age groups (Figure 2A). Grip strength also showed similar results to calf circumference (Figure 2B). On the other hand, the five-time chair stand showed the opposite results to calf circumference, being fastest in the youngest-old group (Figure 2C). ASM was highest in the youngest-old group, followed by the young-old group, and there were no significant differences among the middle-old, old-old, and oldest-old groups (Figure 2D).

In detail, the calf circumference showed differences between the youngest-old and young-old groups, with a decrease of −8.58%. Similarly, there was a −8.40% difference between the young-old and middle-old groups, a −11.05% difference between the middle-old and old-old groups, and a −9.27% difference between the old-old and oldest-old groups. These differences were statistically significant (Z = 140.957, $p = 0.001$, $\eta^2 = 0.357$). Grip strength exhibited variations across age groups: there was a −20.36% decline between the youngest-old and young-old groups, a −9.32% difference between the young-old and middle-old groups, a −21.01% difference between the middle-old and old-old groups, and a −16.72% difference between the old-old and oldest-old groups. These disparities were statistically significant (Z = 142.088, $p = 0.001$, $\eta^2 = 0.348$). The performance in the five-time chair stand test displayed variations among age groups: there was a 32.49% increase between the youngest-old and young-old groups, a 56.81% difference between the young-old and middle-old groups, an 18.95% difference between the middle-old and old-old groups, and a 27.75% difference between the old-old and oldest-old groups. These differences were statistically significant (Z = 319.322, $p = 0.001$, $\eta^2 = 0.723$). The ASM exhibited disparities among age groups: there was a −26.62% decrease between the youngest-old and young-old groups, a −15.19% difference between the young-old and middle-old groups, a 0.33% difference between the middle-old and old-old groups, and a −7.08% difference between the old-old and oldest-old groups. These differences were statistically significant (Z = 176.718, $p = 0.001, \eta^2 = 0.488$).

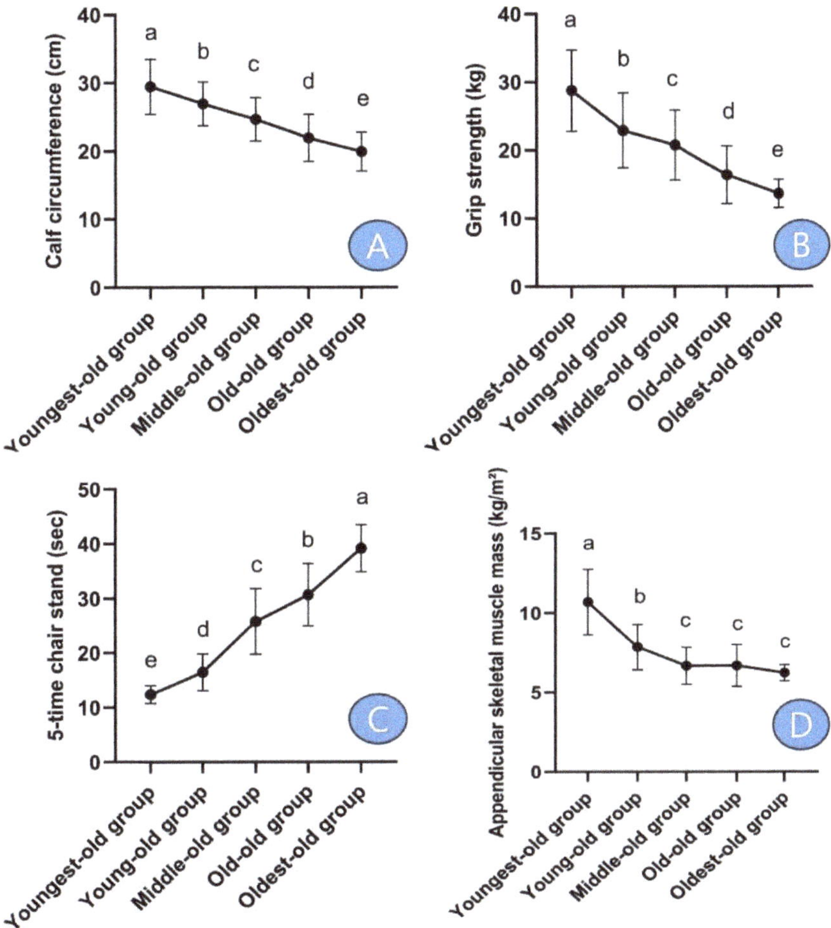

Figure 2. Post-hoc results of sarcopenic factors. Symbols [a, b, c, d,] and [e] represent post hoc results from the Scheffe test. (**A**) means 'calf circumference', (**B**) means 'grip strength', (**C**) means '5-time chair stand', and (**D**) means 'appendicular skeletal muscle mass'.

3.3. Comparative Results of Lymphocyte Subsets

As depicted in Table 3, the white blood cell count, serving as an indicator of immune cells, did not exhibit noteworthy variances between the groups. Nevertheless, significant distinctions were observed among the groups in relation to NK cells, as well as CD3, CD4, and CD8 T cell markers. Additionally, there were discernible variations in CD19, a marker for B cells, across the groups. When analyzing NK cells with age as a covariate, a significant difference was observed among the groups (F = 5.630; $p = 0.001$). Similarly, when analyzing CD3 T cells with age as a covariate, a significant difference was found among the groups (F = 3.796; $p = 0.005$), and CD4 T cells also exhibited a significant difference among the groups (F = 7.688; p = 0.001). Similarly, CD8 T cells (F = 7.947; p = 0.001) and CD9 B cells (F = 9.565; p = 0.001) also displayed significant differences among the groups.

Table 3. Age-dependent disparities in lymphocyte subsets.

	Groups					p	η^2
	Youngest-Old (n = 68)	Young-Old (n = 168)	Middle-Old (n = 127)	Old-Old (n = 46)	Oldest-Old (n = 19)		
WBC ($\times 10^3$ cells/μL)	5.99 $\pm$ 0.91	5.88 $\pm$ 0.71	5.71 $\pm$ 0.84	5.78 $\pm$ 0.82	5.80 $\pm$ 0.66	0.277	0.016
NK cells (%)	23.26 $\pm$ 4.67	17.92 $\pm$ 6.18	14.02 $\pm$ 5.12	10.33 $\pm$ 3.15	9.37 $\pm$ 3.02	0.001	0.371
CD3 T cells (%)	62.05 $\pm$ 8.76	59.52 $\pm$ 8.23	55.27 $\pm$ 7.16	53.74 $\pm$ 6.83	47.31 $\pm$ 7.46	0.001	0.170
CD4 T cells (%)	40.17 $\pm$ 7.60	38.94 $\pm$ 8.24	33.04 $\pm$ 6.56	27.93 $\pm$ 4.62	27.03 $\pm$ 3.65	0.001	0.274
CD8 T cells (%)	27.06 $\pm$ 6.41	22.07 $\pm$ 5.51	19.51 $\pm$ 4.46	15.68 $\pm$ 3.02	13.82 $\pm$ 1.28	0.001	0.328
CD19 B cells (%)	13.71 $\pm$ 2.83	13.18 $\pm$ 3.35	10.54 $\pm$ 2.78	9.56 $\pm$ 2.60	7.98 $\pm$ 2.07	0.001	0.253

All data represent mean $\pm$ standard deviation. WBC, white blood cells; NK, natural killer; CD, cluster differentiation.

Following the post hoc test, it was established that NK cells were most abundant within the youngest-old group. Notably, there was no discernible distinction in NK cell levels between the young-old and middle-old groups. Subsequently, a gradual decrease in NK cell levels was evident in the old-old and oldest-old groups (Figure 3). In detail, the NK cells showed variations across age groups: there was a -22.95% decline between the youngest-old and young-old groups, a -21.79% difference between the young-old and middle-old groups, a -26.27% difference between the middle-old and old-old groups, and a -9.28% difference between the old-old and oldest-old groups. These differences were statistically significant (Z = 159.849, p = 0.001, η^2 = 0.371).

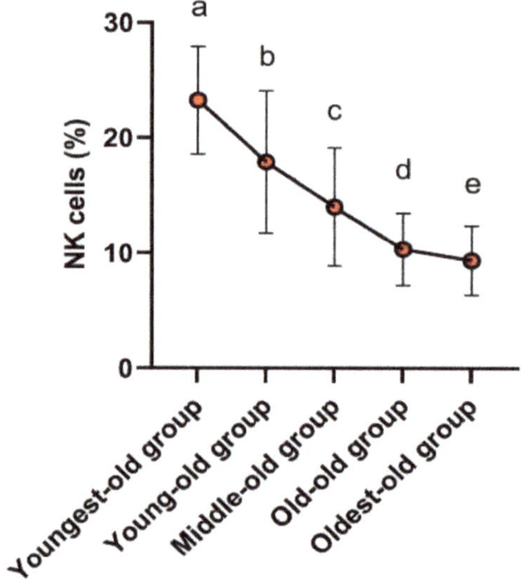

Figure 3. Post-hoc results of innate immune cells. Symbols [a, b, c, d,] and [e] represent post hoc results from the Scheffe test.

The outcomes of the post hoc test revealed that CD3 T cells exhibited their highest levels in both the youngest-old and young-old groups. Subsequently, the middle-old and old-old groups followed suit with slightly lower levels. The lowest levels of CD3 T cells were observed in the oldest-old group (Figure 4A). This tendency was similarly observed in CD4 T cells (Figure 4B), CD8 T cells (Figure 4C), and CD19 B cells (Figure 4D).

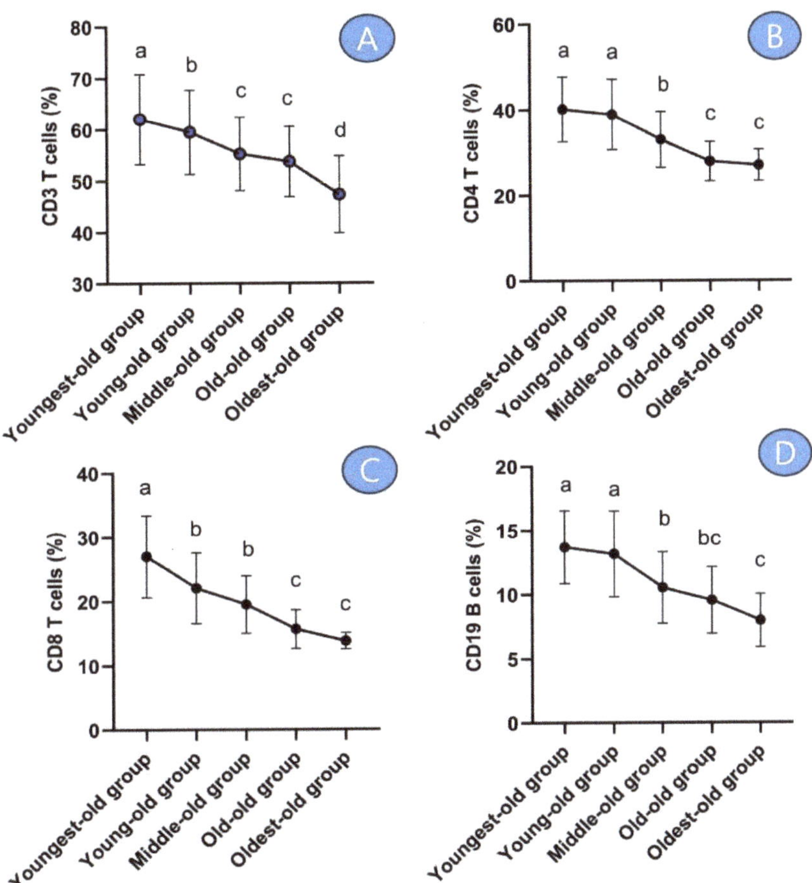

Figure 4. Post-hoc results of adaptive immune cells. Symbols [a, b, c] and [d] represent post hoc results from the Scheffe test. (**A**) means 'CD3 T cells', (**B**) means 'CD4 T cells', (**C**) means 'CD8 T cells', and (**D**) means 'CD19 B cells'.

In detail, the CD3 T cells displayed differences across age groups: there was a −4.07% decrease between the youngest-old and young-old groups, a −7.13% difference between the young-old and middle-old groups, a −2.78% difference between the middle-old and old-old groups, and a −11.95% difference between the old-old and oldest-old groups. These disparities were statistically significant (Z = 68.006, p = 0.001, η^2 = 0.170). The CD4 T cells exhibited disparities among age groups: there was a −3.08% decrease between the youngest-old and young-old groups, a −15.15% difference between the young-old and middle-old groups, a −15.45% difference between the middle-old and old-old groups, and a −3.24% difference between the old-old and oldest-old groups. These differences were statistically significant (Z = 119.355, p = 0.001, η^2 = 0.274). The CD8 T cells displayed differences across age groups: there was a −18.43% decrease between the youngest-old and young-old groups, an −11.61% difference between the young-old and middle-old groups, a −19.64% difference between the middle-old and old-old groups, and an −11.83% difference between the old-old and oldest-old groups. These differences were statistically significant (Z = 135.165, p = 0.001, η^2 = 0.328). The CD19 B cells showed differences across age groups: there was a −3.89% decrease between the youngest-old and young-old groups, a −20.04% difference between the young-old and middle-old groups, a −9.26% difference between the middle-old and old-old groups, and a -16.55% difference between the old-old and oldest-old groups. These differences were statistically significant ($Z = 112.429$, $p = 0.001$, $\eta^2 = 0.253$).

4. Discussion

This study effectively identified distinct alterations in both sarcopenia-related factors and immune cell functionalities when individuals aged 61 to 85 were categorized into five-year groups. The findings revealed that lean mass, as measured by BIA, displayed no significant differences among the youngest-old, young-old, middle-old, and old-old groups. However, a notable decrease was evident after the age of 81 and above. Conversely, fat mass exhibited an increase in all groups older than the middle-old group. On a different note, concerning clinical conditions, there was a noticeable decline in the health status of individuals as age advanced. Specifically, hypertension, arthritis, and low back pain exhibited clear signs of deterioration. Furthermore, it was verified that the daily physical activity levels of the elderly progressively declined with advancing age.

Among the various factors associated with sarcopenia that were examined in this study, both calf circumference and the ability to perform the five-time chair stand consistently exhibited a decline with each successive 5-year age increment. Conversely, grip strength demonstrated a steady reduction after the group reached the age of 66 years or older. Meanwhile, there was a general tendency for ASM levels to decrease as age advanced. Notably, no significant distinction was found between the middle-old and old-old groups in terms of ASM. However, a significant and rapid decrease in ASM became evident after reaching the age of 81.

The prevalence of sarcopenia varies based on the way it is defined in the literature. According to different definitions, the occurrence rate is noted as 5–13% in individuals aged 60–70, whereas it can range from 11% to as high as 50% in those who are over 80 years old [22]. According to the EWGSOP, sarcopenia is described as a prevalent, complex, and costly health condition in the elderly population. Additionally, this geriatric syndrome is attributed to the incomplete interaction between diseases and aging across various systems, resulting in the manifestation of a constellation of signs and symptoms [8,23]. As observed in this study, an increase in age was associated with a clinically deteriorating phenomenon, and the worsening of chronic conditions and physical activity levels aligned with several preceding studies [18,19].

Multiple mechanisms potentially contribute to the initiation and advancement of sarcopenia [24]. These mechanisms encompass various factors, including protein synthesis, proteolysis, neuromuscular integrity, and muscle fat content. In individuals with sarcopenia, a combination of these mechanisms could play a role, with their relative impacts potentially changing over time. Identifying these mechanisms and their root causes is anticipated to aid in the development of intervention studies aiming to address one or more fundamental mechanisms. The concept of staging sarcopenia, which indicates the level of severity of the condition, offers valuable guidance for clinical management. EWGSOP proposes a conceptual staging framework comprising "pre-sarcopenia", "sarcopenia", and "severe sarcopenia [25]". These categorizations imply that the more they are tailored to specific age groups, the more precise they become. The pre-sarcopenia stage is characterized by reduced muscle mass without a noticeable impact on muscle strength or physical performance. This stage can only be identified accurately using methods that measure muscle mass in comparison to established standard populations. The "sarcopenia" stage involves both low muscle mass and either low muscle strength or poor physical performance. The "severe sarcopenia" stage is diagnosed when all three criteria of the definition are met, namely, low muscle mass, low muscle strength, and poor physical performance [9]. Recognizing the different stages of sarcopenia could aid in selecting appropriate treatments and setting realistic recovery objectives. Staging might also facilitate the design of research studies that focus on specific stages or changes in stages over time. Comparing the findings of this study with those of previous research, it can generally be observed that before the age of 70, individuals are in a phase of pre-sarcopenia, and between the ages of 70 and 80, they seem to enter the stage of sarcopenia. However, from the age of 81 onwards, it could be interpreted that individuals transition into a more severe stage of sarcopenia characterized by significant muscle loss.

Similar to the changes in factors contributing to muscle loss and its progression, this study observed that the immune status of older individuals tends to worsen or decline in function as age increases [12]. The immune function plays a vital role as one of the body's defense systems, protecting the body from invading bacteria and antigens [26]. The proper functioning of the immune system within the body is determined, to a large extent, by the presence of a sufficient number of robust immune cells [27]. In this study, the average cell count of white blood cells did not show significant differences with increasing age. However, innate immune cells (NK cells) and adaptive immune cells exhibited a tendency to decrease with age. These changes in the immune system appear to undergo a distinct shift around the age of 70. This age group corresponds to the middle-old cohort examined in this study, specifically individuals aged 71–75 years old. These results are consistent with studies indicating a reduction in the number of peripheral blood lymphocytes with aging [28,29], suggesting that lymphocyte subtypes are sensitive to the effects of aging. It should be noted, however, that there are also studies suggesting no change in lymphocyte count with age [28], indicating that further research is needed in the future. Consequently, age-related dysregulation and senescence of the immune system could potentially contribute to the advancement and deterioration of sarcopenia [29,30]. The immune system may participate in regulating skeletal muscle growth and regeneration in instances of acute and chronic muscle injuries [19,31]. Therefore, it is plausible to suggest that changes in the immune system associated with aging might play a significant role in the progression of sarcopenia.

In essence, this study demonstrated a consistent trend of decline in both sarcopenia-related factors and immune cell functionality as age progressed beyond 61 years. Furthermore, these findings were notably associated with the reduction in lean mass observed in clinical characteristics. It is reasonable to anticipate that implementing strategies to prevent the decline in muscle mass as individuals enter old age after the age of 61 can contribute to maintaining a high quality of life and preserving immune cell function. For example, as individuals age, it is essential to examine their susceptibility to easily occurring infections and their responsiveness to vaccinations [32]. Additionally, it is crucial to identify and manage the underlying causes of various age-related diseases, such as obesity, hypertension, atherosclerosis, osteoporosis, diabetes, and cancer [33,34]. Moreover, maintaining regular exercise habits [18], adopting a healthy diet, ensuring proper sleep patterns, and managing stress are all important measures to prevent the deterioration of immune cell function [17,35]. Ultimately, this study discovered that with advancing age, muscle mass decreases, evaluation parameters related to muscle loss factors worsen, and consequently, immune cell function deteriorates. Furthermore, it was determined that these findings were more pronounced when age was categorized into 5-year intervals. However, this study had a small sample size due to the limited number of older adults in the studied community in Korea. Furthermore, the fact that only elderly individuals residing in Korea were included as the study population could serve as a limitation.

5. Conclusions

In this study, when age was categorized into 5-year intervals, the researcher was able to observe a distinct pattern of muscle atrophy, and at the same time, the researcher confirmed a decline in the functions of immune cells. To prevent muscle atrophy in the elderly and observe declines in immune cell function, it is advisable to use 5-year age intervals, and through this approach, we can anticipate a higher quality of life.

Author Contributions: S.-J.H. conceived the idea. S.-J.H. developed the background and performed the calibration of different devices used in the tests. S.-J.H. verified the methods section. S.-J.H. wrote the manuscript and contributed to the final version of the manuscript. Y.-S.J. contributed to the interpretation of the results and data analysis, and he wrote and revised the manuscript. All authors have read and agreed to the published version of the manuscript.

Funding: This research received no external funding.

Institutional Review Board Statement: The study was conducted according to the guidelines of the Declaration of Helsinki and approved by the Institutional Review Board of Hanseo University (HS23-08-02, 2 November 2022 to 1 November 2023).

Informed Consent Statement: Informed consent was obtained from all subjects involved in the study.

Data Availability Statement: Not applicable.

Acknowledgments: The author wishes to thank the participants of this study.

Conflicts of Interest: The authors declare no conflict of interest.

References

1. Vogelsang, E.M.; Raymo, J.M.; Liang, J.; Kobayashi, E.; Fukaya, T. Population Aging and Health Trajectories at Older Ages. *J. Gerontol. B Psychol. Sci. Soc. Sci.* **2019**, *74*, 1245–1255. [CrossRef]
2. Schulz, R.; Martire, L.M. Family caregiving of persons with dementia: Prevalence, health effects, and support strategies. *Am. J. Geriatr. Psychiatry* **2004**, *12*, 240–249. [CrossRef]
3. Ory, M.; Hoffman, M.K.; Hawkins, M.; Sanner, B.; Mockenhaupt, R. Challenging aging stereotypes: Strategies for creating a more active society. *Am. J. Prev. Med.* **2003**, *25* (Suppl. S2), 164–171. [CrossRef]
4. Boeckxstaens, P.; De Graaf, P. Primary care and care for older persons: Position paper of the European Forum for Primary Care. *Qual. Prim. Care* **2011**, *19*, 369–389.
5. Dear, B.F.; Scott, A.J.; Fogliati, R.; Gandy, M.; Karin, E.; Dudeney, J.; Nielssen, O.; McDonald, S.; Heriseanu, A.I.; Bisby, M.A.; et al. The Chronic Conditions Course: A Randomised Controlled Trial of an Internet-Delivered Transdiagnostic Psychological Intervention for People with Chronic Health Conditions. *Psychother. Psychosom.* **2022**, *91*, 265–276. [CrossRef]
6. Mijnarends, D.M.; Luiking, Y.C.; Halfens, R.J.G.; Evers, S.M.A.A.; Lenaerts, E.L.A.; Verlaan, S.; Wallace, M.; Schols, J.M.G.A.; Meijers, J.M.M. Muscle, Health and Costs: A Glance at their Relationship. *J. Nutr. Health Aging* **2018**, *22*, 766–773. [CrossRef]
7. Chen, L.K.; Woo, J.; Assantachai, P.; Auyeung, T.W.; Chou, M.Y.; Iijima, K.; Jang, H.C.; Kang, L.; Kim, M.; Kim, S.; et al. Asian Working Group for Sarcopenia: 2019 consensus update on sarcopenia diagnosis and treatment. *J. Am. Med. Dir. Assoc.* **2020**, *21*, 300–307e2. [CrossRef]
8. Cruz-Jentoft, A.J.; Landi, F.; Topinková, E.; Michel, J.P. Understanding sarcopenia as a geriatric syndrome. *Curr. Opin. Clin. Nutr. Metab Care* **2010**, *13*, 1–7. [CrossRef]
9. Cruz-Jentoft, A.J.; Bahat, G.; Bauer, J.; Boirie, Y.; Bruyère, O.; Cederholm, T.; Cooper, C.; Landi, F.; Rolland, Y.; Sayer, A.A.; et al. Writing Group for the European Working Group on Sarcopenia in Older People 2 (EWGSOP2), and the Extended Group for EWGSOP2. Sarcopenia: Revised European consensus on definition and diagnosis. *Age Ageing* **2019**, *48*, 16–31. [CrossRef]
10. Cosquéric, G.; Sebag, A.; Ducolombier, C.; Thomas, C.; Piette, F.; Weill-Engerer, S. Sarcopenia is predictive of nosocomial infection in care of the elderly. *Br. J. Nutr.* **2006**, *96*, 895–901. [CrossRef]
11. Tidball, J.G.; Welc, S.S.; Wehling-Henricks, M. Immunobiology of Inherited Muscular Dystrophies. *Compr. Physiol.* **2018**, *8*, 1313–1356. [CrossRef]
12. Nelke, C.; Dziewas, R.; Minnerup, J.; Meuth, S.G.; Ruck, T. Skeletal muscle as potential central link between sarcopenia and immune senescence. *EBioMedicine* **2019**, *49*, 381–388. [CrossRef]
13. Thériault, M.E.; Paré, M.È.; Maltais, F.; Debigaré, R. Satellite cells senescence in limb muscle of severe patients with COPD. *PLoS ONE* **2012**, *7*, e39124. [CrossRef]
14. Cheng, H.L. A simple, Easy-to-Use Spreadsheet for Automatic Scoring of the International Physical Activity Questionnaire (IPAQ) Short Form. 2016. Available online: https://www.researchgate.net/publication/310953872_A_simple_easy-touse_spreadsheet_for_automatic_scoring_of_the_International_Physical_Activity_Questionnaire_IPAQ_Short_Form (accessed on 6 June 2023).
15. He, H.; Pan, L.; Wang, D.; Liu, F.; Du, J.; Pa, L.; Wang, X.; Cui, Z.; Ren, X.; Wang, H.; et al. Normative values of hand grip strength in a large unselected Chinese population: Evidence from the China National Health Survey. *J. Cachexia Sarcopenia Muscle* **2023**, *14*, 1312–1321. [CrossRef]
16. Wu, X.; Li, X.; Xu, M.; Zhang, Z.; He, L.; Li, Y. Sarcopenia prevalence and associated factors among older Chinese population: Findings from the China Health and Retirement Longitudinal Study. *PLoS ONE* **2021**, *16*, e0247617. [CrossRef]
17. Heo, S.J.; Park, S.K.; Jee, Y.S. Effects of phytoncide on immune cells and psychological stress of gynecological cancer survivors: Randomized controlled trials. *J. Exerc. Rehabil.* **2023**, *19*, 170–180. [CrossRef]
18. Park, S.; Park, S.K.; Jee, Y.S. Moderate- to fast-walking improves immunocytes through a positive change of muscle contractility in old women: A pilot study. *J. Exerc. Rehabil.* **2023**, *19*, 45–56. [CrossRef]
19. Jee, Y.S. Influences of acute and/or chronic exercise on human immunity: Third series of scientific evidence. *J. Exerc. Rehabil.* **2020**, *16*, 205–206. [CrossRef]

20. Franceschi, C.; Garagnani, P.; Parini, P.; Giuliani, C.; Santoro, A. Inflammaging: A new immune-metabolic viewpoint for age-related diseases. *Nat. Rev. Endocrinol.* **2018**, *14*, 576–590. [CrossRef]
21. Cohen, J. A power primer. *Psychol. Bull.* **1992**, *112*, 155–159. [CrossRef]
22. Rolland, Y.; Czerwinski, S.; Van Kan, G.A.; Morley, J.E.; Cesari, M.; Onder, G.; Woo, J.; Baumgartner, R.; Pillard, F.; Boirie, Y.; et al. Sarcopenia: Its assessment, etiology, pathogenesis, consequences and future perspectives. *J. Nutr. Health Aging* **2008**, *12*, 433–450. [CrossRef] [PubMed]
23. Cruz-Jentoft, A.J.; Baeyens, J.P.; Bauer, J.M.; Boirie, Y.; Cederholm, T.; Landi, F.; Martin, F.C.; Michel, J.P.; Rolland, Y.; Schneider, S.M.; et al. Sarcopenia: European consensus on definition and diagnosis: Report of the European Working Group on Sarcopenia in Older People. *Age Ageing* **2010**, *39*, 412–423. [CrossRef] [PubMed]
24. Gallagher, D.; Visser, M.; De Meersman, R.E.; Sepúlveda, D.; Baumgartner, R.N.; Pierson, R.N.; Harris, T.; Heymsfield, S.B. Appendicular skeletal muscle mass: Effects of age, gender, and ethnicity. *J. Appl. Physiol.* **1997**, *83*, 229–239. [CrossRef] [PubMed]
25. D'Onofrio, G.; Cseri, J. (Eds.) *Frailty and Sarcopenia: Recent Evidence and New Perspectives*; Books on Demand GmbH: Norderstedt, Germany, 2022.
26. Weiskopf, D.; Weinberger, B.; Grubeck-Loebenstein, B. The aging of the immune system. *Transpl. Int.* **2009**, *22*, 1041–1050. [CrossRef]
27. Vitlic, A.; Lord, J.M.; Phillips, A.C. Stress, ageing and their influence on functional, cellular and molecular aspects of the immune system. *Age* **2014**, *36*, 9631. [CrossRef]
28. Olsson, J.; Wikby, A.; Johansson, B.; Löfgren, S.; Nilsson, B.O.; Ferguson, F.G. Age-related change in peripheral blood T-lymphocyte subpopulations and cytomegalovirus infection in the very old: The Swedish longitudinal OCTO immune study. *Mech. Ageing Dev.* **2000**, *121*, 187–201. [CrossRef]
29. Freitas, G.R.R.; da Luz Fernandes, M.; Agena, F.; Jaluul, O.; Silva, S.C.; Lemos, F.B.C.; Coelho, V.; Elias, D.N.; Galante, N.Z. Aging and End Stage Renal Disease Cause A Decrease in Absolute Circulating Lymphocyte Counts with A Shift to A Memory Profile and Diverge in Treg Population. *Aging Dis.* **2019**, *10*, 49–61. [CrossRef]
30. Zanni, F.; Vescovini, R.; Biasini, C.; Fagnoni, F.; Zanlari, L.; Telera, A.; Di Pede, P.; Passeri, G.; Pedrazzoni, M.; Passeri, M.; et al. Marked increase with age of type 1 cytokines within memory and effector/cytotoxic CD8+ T cells in humans: A contribution to understand the relationship between inflammation and immunosenescence. *Exp. Gerontol.* **2003**, *38*, 981–987. [CrossRef]
31. Yang, W.; Hu, P. Skeletal muscle regeneration is modulated by inflammation. *J. Orthop. Translat.* **2018**, *13*, 25–32. [CrossRef]
32. Gavazzi, G.; Herrmann, F.; Krause, K.H. Aging and infectious diseases in the developing world. *Clin. Infect. Dis.* **2004**, *39*, 83–91. [CrossRef]
33. Griffin, W.S. Inflammation and neurodegenerative diseases. *Am. J. Clin. Nutr.* **2006**, *83*, 470S–474S. [CrossRef] [PubMed]
34. Castelo-Branco, C.; Soveral, I. The immune system and aging: A review. *Gynecol. Endocrinol.* **2014**, *30*, 16–22. [CrossRef] [PubMed]
35. Walsh, N.P.; Gleeson, M.; Pyne, D.B.; Nieman, D.C.; Dhabhar, F.S.; Shephard, R.J.; Oliver, S.J.; Bermon, S.; Kajeniene, A. Position statement. Part two: Maintaining immune health. *Exerc. Immunol. Rev.* **2011**, *17*, 64–103. [PubMed]

Article

Delayed Surgical Treatment in Patients with Chronic Carpal Tunnel Syndrome Is Still Effective in the Improvement of Hand Function

Marta Twardowska [1,*], Piotr Czarnecki [1], Marta Jokiel [1], Ewa Bręborowicz [1], Juliusz Huber [2] and Leszek Romanowski [1]

[1] Department of Traumatology, Orthopaedics and Hand Surgery, Poznan University of Medical Sciences, 28 Czerwca 1956 r. Street, No. 135/147, 61-545 Poznan, Poland; piotr_czarnecki@tlen.pl (P.C.); marta.jokiel@gmail.com (M.J.); ewabreborowicz@gmail.com (E.B.); lromanowski@orsk.pl (L.R.)

[2] Department of Pathophysiology of Locomotor Organs, Poznan University of Medical Sciences, 28 Czerwca 1956 r. Street, No. 135/147, 61-545 Poznan, Poland; juliusz.huber@ump.edu.pl

* Correspondence: marcik.twardowska@gmail.com

Abstract: *Background and Objectives*: Severe carpal tunnel syndrome (CTS) is the most common compression neuropathy in the upper extremities treated conservatively; later, when advanced, CTS is treated mostly surgically. The most prevalent symptoms comprise numbness, as well as sensation loss in the thumb, index, and middle finger, and thenar muscle strength loss, resulting in impaired daily functioning for patients. Data on the results of CTS treatment in patients with delayed surgical intervention are scarce. The aim of this study was to determine the postoperative results of chronic carpal tunnel syndrome treatment in patients with symptoms lasting for at least 5 years. *Materials and Methods*: A total of 86 patients (69 females, 17 males) with a mean age of 58 years reporting symptoms of CTS for at least 5 years (mean: 8.5 years) were prospectively studied. The average follow-up time was 33 months. All patients underwent the surgical open decompression of the median nerve at the wrist. A preoperative observation was composed of an interview and a clinical examination. The subjects completed the DASH (the Disabilities of the Arm, Shoulder, and Hand), PRWE (Patient-Rated Wrist Evaluation), and self-report questionnaires. Global grip strength, sensory discrimination, characteristic symptoms of CTS, and thenar muscle atrophy were examined. Postoperatively, clinical and functional examinations were repeated, and patients expressed their opinions by completing a BCTQ (Boston Carpal Tunnel Syndrome Questionnaire). *Results*: We found improvements in daily activities and hand function postoperatively. Overall, 88% of patients were satisfied with the outcome of surgery. DASH scores decreased after surgery from 44.82 to 14.12 at $p < 0.001$. PRWE questionnaire scores decreased from 53.34 to 15.19 at $p < 0.001$. The mean score of the BCTQ on the scale regarding the severity of symptoms was 1.48 and 1.62 on the scale regarding function after surgery. No significant differences were found in the scores between the male and female groups or between age groups ($p > 0.05$). A significant increase in global grip strength from 16.61 kg to 21.91 kg was observed postoperatively at $p < 0.001$. No significant difference was detected in the measurement of sensory discrimination (6.02 vs. 5.44). In most of the examined patients, night numbness and wrist pain subsided after surgery at $p < 0.001$. Thenar muscle atrophy diminished after surgery at $p < 0.001$. *Conclusions*: Most patients were satisfied with the results of CTS surgery regarding the open decompression of the median nerve even after 5 years of ineffective conservative treatment. Significant improvement of the hand function was confirmed in the functional studies.

Keywords: chronic carpal tunnel syndrome; median nerve neuropathy; delayed surgery; effectiveness of treatment; functional tests; questionnaires evaluation

Citation: Twardowska, M.; Czarnecki, P.; Jokiel, M.; Bręborowicz, E.; Huber, J.; Romanowski, L. Delayed Surgical Treatment in Patients with Chronic Carpal Tunnel Syndrome Is Still Effective in the Improvement of Hand Function. *Medicina* **2023**, *59*, 1404. https://doi.org/10.3390/medicina59081404

Academic Editors: Milica Lazovic and Dejan Nikolić

Received: 27 June 2023
Revised: 27 July 2023
Accepted: 29 July 2023
Published: 31 July 2023

1. Introduction

Carpal tunnel syndrome (CTS) is the most common compression neuropathy in the upper extremity, affecting 2–4% of the population [1]. Compression of the nerve causes numbness in the middle finger, index finger, and thumb, diminished sensation, and atrophy of the thenar muscle, which leads to deterioration in the patient's daily living activities [2]. Finger numbness often occurs at night, causing difficulties in sleeping. Risk factors for CTS may include rheumatoid diseases, diabetes, bone fractures, using a computer, obesity, hypothyroidism, and hormone replacement therapy [2–4]. However, in most cases, the cause is unexplained, leading to idiopathic CTS [5].

A detailed patient history and clinical examination are required for CTS diagnosis. Sometimes, they are sufficient for a correct diagnosis confirming CTS [5,6]; additional methods may include an X-ray, ultrasound, and nerve conduction studies (electroneurography or electromyography) [7–10]. These tests can help rule out other causes of patient complaints.

Treatment for CTS may be non-operative initially, involving physiotherapy (e.g., manual therapy and neuromobilization), the use of orthoses, or corticosteroid injections, although surgical treatment is indicated for patients with advanced disease severity [5]. Possible surgical techniques are open surgery or endoscopic techniques [6]. Both methods involve transection of the flexor retinaculum, thus decompressing the median nerve [11]. There are randomized controlled trials and numerous observational studies in the literature which have confirmed the differences and similarities in postoperative results between open and endoscopic methods of median nerve decompression yield [11,12].

There are scarce data on the results of CTS treatment in patients with delayed surgical intervention. The rationale of this study was to provide data that could help to discuss the prognosis and results of treatment in these patients. The aim of this study was to determine the postoperative clinical and functional results of surgery in chronic carpal tunnel syndrome treatment cases, i.e., those patients for whom the above symptoms have lasted for at least 5 years, which is a novelty in the literature on the topic. We used a set of both clinical and questionnaire tools which have never been reported before. The null hypothesis is that CTS treatment in patients with delayed surgical intervention does not bring the significant improvement. The secondary aim of the study was to age-wise and gender-wise evaluate the results with the hypothesis that age and gender do not influence the results of the treatment.

2. Materials and Methods

2.1. Subjects, Study Design, and Clinical Evaluation

This study was performed on 114 patients diagnosed with chronic carpal tunnel syndrome (CCTS) who underwent surgery in the Traumatology, Orthopedics, and Hand Surgery Department Poznan University of Medical Sciences, Poland, from 2014 to 2018. The study was approved by the Bioethics Committee at the Poznan University of Medical Sciences (Resolution no. 32/15).

The diagnosis of CTS was confirmed by a medical team (with at least 5 experienced hand surgeons) during clinical meetings. Patient history was analyzed, with patients exhibiting characteristic symptoms previously described in the literature. The Phalen test and Tinnel–Hoffman sign had to be positive. We considered these two clinical tests reliable enough to confirm the symptoms of the carpal tunnel syndrome and did not perform any other clinical tests like the reversed Phalen or Durkan's test. The combination of the Tinel–Hoffmann sign together with Phalen test proved to be of high sensitivity and specificity—similar to the Phalen and Durkan test combination [13,14].

We considered the minimum duration of symptoms before surgery for 5 years because such a period causes permanent damage to the nerve, so the benefits of surgical treatment can potentially be less satisfactory than in patients treated earlier.

The study group consisted of patients who, despite the use of various conservative methods (involving physiotherapy, e.g., manual therapy and neuromobilization, the use of orthoses, or corticosteroid injection), still reported severe symptoms of CTS and therefore qualified for surgical treatment. During the study, electrophysiological test results were not provided because this was not the purpose of the study. In our department, imaging studies and nerve conduction studies are not obligatory for carpal tunnel diagnosis. These modalities are used in revision cases or in multiple crush syndromes [15,16], which were excluded from this study.

After the diagnosis was verified, patients qualified for open median nerve decompression surgery. All patients underwent the surgery in the same hospital according to the department's approved procedure. Subsequently, all patients underwent a follow-up examination in 2017–2019, which was conducted a minimum of 12 months after surgery. At the end of 2019, some of the patients might have already been the first victims infected with COVID-19, which is known for causing polyneuropathies; therefore, we accepted this year as the ending point of the follow-up [17–19].

The preoperative examination included a group of 114 patients, but some of them did not meet the inclusion or exclusion criteria or did not attend the follow-up examination after surgery because of COVID-related problems (Figure 1). Finally, 86 patients qualified for the second stage of the study (Table 1). In cases of bilateral CTS, only one hand with more severe symptoms was operated on and evaluated in the period of follow-up.

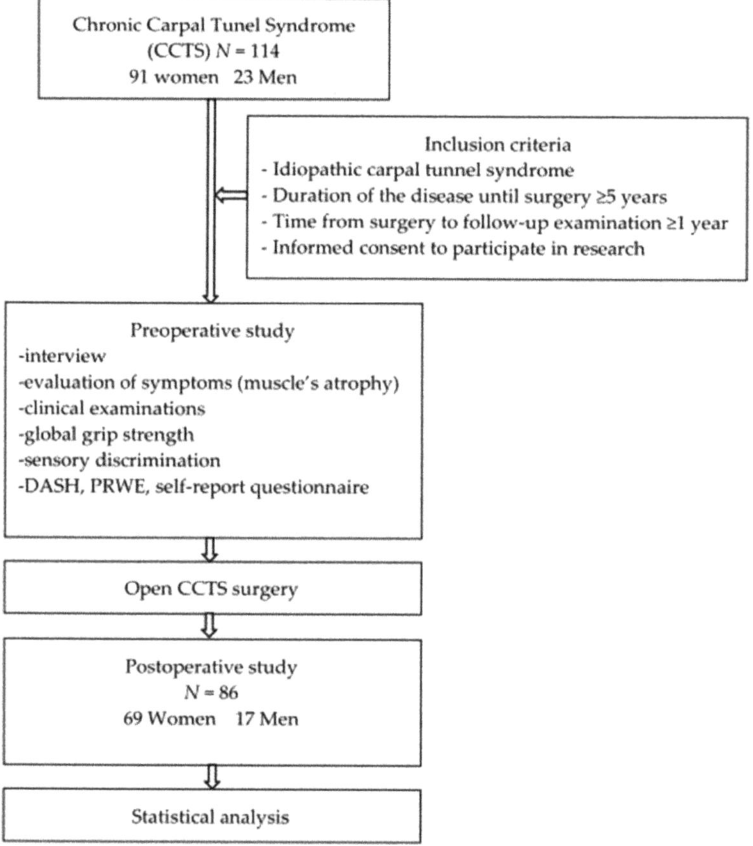

Figure 1. Flow chart detailing the research procedure.

Table 1. The mean age of the patients studied, the duration of the disease (expressed in years), and the time from surgery to follow-up (expressed in months).

	Mean	**Minimum**	**Maximum**	**SD**
Age (N = 86)	57.76	34	85	8.71
Disease duration	8.47	5	20	3.08
Follow-up	33.48	12	65	7.009

Abbreviations: N, group size; SD, standard deviation.

However, the long time between the decision for surgical treatment and the performance of surgery was not due to health system organizational reasons. The most common reasons for delaying the surgical CTS procedure reported by the patients were fear of the procedure, waiting for improvement from conservative treatment, and the lack of a precise diagnosis of the disease.

This study was prospective and included two stages. The first involved a preoperative examination. Patients underwent a clinical examination, and global grip strength and sensory abnormalities were evaluated.

The inclusion criteria were as follows:

- Idiopathic carpal tunnel syndrome;
- Duration of the disease until surgery $\geq$5 years;
- Time from surgery to follow-up examination $\geq$1 year;
- Informed consent to participate in research.

The exclusion criteria were as follows:

- Diagnosed diseases of the cervical spine, e.g., consequences of mechanical injuries, disc-root conflicts, or degenerative disease of the spine during treatment;
- Previous injuries within the examined upper extremities and diseases that may affect the function of the median nerve, e.g., bone fractures, consequences of upper limb ischemia, thoracic outlet syndrome, and shoulder injuries;
- Patients with previous carpal tunnel syndrome surgeries or multiple injections of corticosteroids,
- Symptoms of other concomitant hand problems like triggering of the fingers, thumb carpometacarpal (CMC) I joint arthritis;
- The presence of polyneuropathy symptoms.

On the basis of medical data obtained from hospital records and the medical history collected during the study, we created a database containing information on the characteristics of symptoms, conservative treatment, type of work, comorbidities, and other relevant information. During the clinical examination, we examined the Phalen and Tinnel–Hoffman signs, and we assessed the thenar muscle mass. The thenar muscle mass was comparatively visually inspected before and after the surgery on a two-stage scale: 1—normal and 0—thenar muscle atrophy. The global grip strength test consisted of measuring force with an electronic dynamometer, i.e., the Biometric LTD Hand Kit. The patient made three attempts to tighten the handle of the dynamometer with the highest possible force, after which the average of the three results was calculated. The test took place in a standing position, with the elbow joint placed against the torso and bent at an angle of 90°; then, the test hand tightened the dynamometer handle. The wrist had to be positioned at 0° or in slight dorsiflexion (approximately 10–20°) to offset the effect of the flexor retinaculum on grip strength. Sensation was measured using a two-point discrimination (2PD) test [20]. It was carried out using the so-called Weber's compass, which allows the fingertips of the examined patient to be touched in two places at the same time. The shortest distance between two points that can be recognized by the patient as two separate points is the measure of discrimination. Under normal conditions, it ranges from 3 to 6 mm. Results above 15 mm refer to an undeterminable resolution of sensation.

2.2. Questionnaires and Surgery

Patients completed the DASH (the Disabilities of the Arm, Shoulder, and Hand) and PRWE (Patient-Rated Wrist Evaluation) questionnaires before the operation. After the operation, in addition to these two questionnaires, they also completed the BTCQ (Boston Carpal Tunnel Syndrome Questionnaire). All questionnaires have been validated in and adapted to Polish [21,22].

The DASH questionnaire consists of 30 questions, of which 21 concern the assessment of limb function, and 6 questions concern symptoms such as pain, tingling, weakness, and stiffness [23]. The next 3 concern interpersonal relationships and self-perception. The questionnaire ranges from a scale of 1–5. To be able to calculate the result of the questionnaire, the respondent must answer at least 27 out of 30 questions. The responses are summed up, and the resulting sum is divided by the number of answers given, obtaining an average answer in the range from 1 to 5. The obtained result is transformed to a 100-point scale by subtracting 1 and multiplying by 25. The above transformation is intended to facilitate comparisons with other scoring questionnaires in the range from 0 to 100. All questions are about complaints in the last week.

The PRWE questionnaire assesses the level of pain and the degree of difficulty caused by specific activities to the patient [24].

It consists of 15 questions. It is divided into 2 parts:

- The first consists of 5 questions about the intensity of hand pain.
- The second consists of 10 hand function assessment questions.

In both parts, the patient evaluates the average severity of symptoms over the last week. The patient marks the average of symptoms on a scale from 0 to 10, where 0 means no pain or no difficulty in performing the given activities, and 10 means the greatest pain experienced or always present pain or impossible activity.

After the examination, patients underwent open decompression surgery, which involved surgical incision of the flexor retinaculum and releasing the median nerve compression. The duration of the surgery, including anesthesia, was 45 min on average. We used a tourniquet during surgery for 30 min, but it was deflated before the closure to check hemostasis. The surgery was performed with the brachial plexus anesthesia routinely utilized in out department during CTS surgery. A longitudinal incision approximately 2–3 cm long was made on the palmar surface above the carpal tunnel, distally to the wrist flexion crease. After skin incision, the palmar aponeurosis was cut, followed by a flexor retinaculum cut in the proximal and distal directions. A complete cut was verified with scissors. Hemostasis was then performed, and a skin suture was placed using absorbable thread (Rapid 3/0 or 4/0-Yavo Poland). All patients underwent the same skin closure method. We did not apply plaster immobilization, only a soft dressing.

The patients were followed up for a minimum of one year. Out of 114 patients, 28 did not participate in follow-up for various individual reasons (e.g., refusal to come for examination, death, other medical conditions). In addition, some data were missing in pre-op and post-op documentation (grip strength and 2PD examination). The second stage of the study included a post-operative follow-up examination of the operated hand, during which, in addition to the above, patients completed the BCTQ. The BCTQ is a dedicated survey for CTS patients [23,25,26]. The questionnaire consists of two parts. The first of them, called the Symptom Severity Scale (SSS), contains 11 questions concerning the frequency and perception of pain by the patient day and night, numbness and tingling of the fingers, as well as difficulties in gripping small objects such as a pen or keys. The second part, called the Functional Status Scale (FSS), consists of 8 questions about the patient's functioning and the degree of difficulty with everyday activities. In each part, the patient selects the answers on a scale of 1–5, where 1 means the least severity of symptoms or no difficulties with a given activity, and 5 means the most severe symptoms or inability to perform activities.

The patients were additionally asked about satisfaction with the result of the operation (yes, no, hard to say), and then there were open-ended questions where the patients could describe their symptoms and whether they changed over time. We also asked if the patient was still working and after what time he or she returned to work.

2.3. Statistical Analysis

On the basis of medical data obtained from hospital records and our own measurements, a database was created. The data were then statistically analyzed using Statistica 13.0; TIBCO Software Inc., Palo Alto, CA, USA (2017). Statistically significant results were considered those in which the significance level (p) was below 0.05.

Tests were selected according to the distribution of the variables. The absence of a normal distribution in the study group was established using the Shapiro–Wilk test.

According to the distribution of parametric and nonparametric tests, the following tests were used:

- Parametric tests, when the variables met the assumptions of a normal distribution: Student's *t*-test for dependent variables, and Student's *t*-test for the independent variables;
- Non-parametric tests when variables did not meet the assumptions of normal distribution: Wilcoxon test for dependent variables, Mann–Whitney U test for independent variables.

When comparing more than two groups at the same time, Kruskal–Wallis nonparametric tests were used in addition to Dunn's post-hoc tests. When analyzing related variables on a nominal scale (in the case of Tinel's and Phalen's characteristic symptoms), the chi-squared test was used.

When the distribution was normal, the mean and standard deviation were used; when the data did not follow a normal distribution, the median and quartile range were used.

For grip strength, the 2PD Wilcoxon Matched Pairs Test was used. For the PRWE, DASH, BCTQ, and age-wise and gender-wise comparisons, the Mann–Whitney U Test was used. Spearman's rank correlation was used for correlations calculations between the duration of symptoms, follow-up, and PRWE, DASH, and BCTQ questionnaires. Kruskal–Wallis tests were used for the analysis of self-reported questionnaire answers in terms of symptom duration.

3. Results

The dominant hand in 93% of patients was the right hand. In 47% of patients, the left hand was operated on; 53% had their right hand operated on. No trigger fingers were reported or detected postoperatively by the patients.

Most of the patients were satisfied with the outcome of the surgery. In total, 74 people (88% of the surveyed subjects) reported that they were satisfied with the result of the surgery, with only five patients reporting that the result of the procedure was not satisfactory.

Before the surgery, 39 subjects (49%) exhibited atrophy of the thenar muscles, and after the operation, atrophy was found in 29 subjects (36%), i.e., a significant improvement (regeneration of the muscle mass) was observed in 13% of the subjects ($p < 0.001$). The time when thenar muscles were regenerated postoperatively was about 24 months on average.

A significant ($p < 0.001$) increase in global grip strength from 16.61 ± 8.47 kg to 21.91 ± 8.26 kg was also noted (Table 2). No significant difference was observed in the measurement of sensory discrimination. Before surgery, patients revealed scores of 6.02 ± 2.51 mm, whereas after surgery, they had scores of 5.44 ± 1.66 (Table 3).

Table 2. Results of the grip strength analysis, expressed in kilograms.

Variable	Group	N	Median	Q_1	Q_3	p
Grip strength before surgery		80	15.58	10.50	20.50	0.000001
Grip strength after surgery		86	22.35	16.37	25.83	
Grip strength before surgery	w	63	14.50	10.20	19.00	0.000001
Grip strength after surgery	w	69	21.20	15.20	24.57	

Table 2. *Cont.*

Variable	Group	N	Median	Q_1	Q_3	p
Grip strength before surgery	m	17	20.70	16.20	34.83	0.102435
Grip strength after surgery	m	17	27.53	23.97	34.27	
Grip strength before surgery	<60 years	45	15.50	10.27	22.07	0.000297
Grip strength after surgery	<60 years	48	24.37	16.97	27.52	
Grip strength before surgery	>60 years	35	15.67	10.67	20.10	0.000786
Grip strength after surgery	>60 years	38	20.95	15.83	23.87	

Abbreviations: N, group size; Q_1, lower quartile; Q_3, upper quartile; w, women; m, men.

Table 3. Results of the analysis of the discrimination of sensation on fingers I–III expressed in millimeters.

Variable	Group	N	Median	Q_1	Q_3	p
2PD before surgery		84	5.38	4.50	7.63	0.073486
2PD after surgery		85	5.00	4.50	6.00	
2PD before surgery	w	67	6.00	4.50	8.00	0.026250
2PD after surgery	w	68	5.00	4.38	6.00	
2PD before surgery	m	17	5.25	4.50	6.50	0.334278
2PD after surgery	m	17	6.00	5.25	6.50	
2PD before surgery	<60 years	47	5.25	4.50	7.50	0.009552
2PD after surgery	<60 years	47	5.00	4.00	5.25	
2PD before surgery	>60 years	37	6.25	4.50	8.00	0.825912
2PD after surgery	>60 years	38	6.00	5.00	6.75	

Abbreviations: N, group size; Q_1, lower quartile; Q_3, upper quartile; w, women; m, men.

Patients with chronic CTS demonstrated significant improvements in performing daily activities and hand function postoperatively (Tables 4–6).

Table 4. Results of the DASH questionnaire analysis.

Variable	Group	N	Median	Q_1	Q_3	p
Score before surgery		86	47.92	35.00	57.50	0.000000
Score after surgery		86	8.33	0.00	23.33	
Score before surgery	w	69	45.00	35.83	57.50	0.000000
Score after surgery	w	69	11.67	0.00	25.00	
Score before surgery	m	17	50.83	35.00	55.83	0.000293
Score after surgery	m	17	2.50	0.00	8.33	
Score before surgery	<60 years	48	46.25	35.00	60.83	0.000000
Score after surgery	<60 years	48	2.50	0.00	21.25	
Score before surgery	>60 years	38	49.58	35.83	56.67	0.000000
Score after surgery	>60 years	38	13.33	0.83	23.33	

Abbreviations: N, group size; Q_1, lower quartile; Q_3, upper quartile; w, women; m, men.

Significantly ($p < 0.001$) lower DASH questionnaire scores were observed in patients after surgery compared with patients before surgery. DASH scores decreased from 44.82 ± 18.14 to 14.12 ± 18.10 (Table 4). PRWE questionnaire scores significantly ($p < 0.001$) decreased from 53.34 ± 21.17 to 15.19 ± 22.45 (Table 5).

Table 5. Results of the analysis of the PRWE questionnaire.

Variable	Group	N	Median	Q$_1$	Q$_3$	*p*
PRWE before surgery		86	55.25	42.50	68.00	0.000000
PRWE after surgery		86	4.50	0.00	23.50	
PRWE before surgery	w	69	57.00	43.50	68.00	0.000000
PRWE after surgery	w	69	5.00	0.00	27.00	
PRWE before surgery	m	17	44.50	33.00	66.00	0.000293
PRWE after surgery	m	17	1.00	0.00	10.00	
PRWE before surgery	<60 years	48	57.00	45.25	68.00	0.000000
PRWE after surgery	<60 years	48	2.50	0.00	12.75	
PRWE before surgery	>60 years	38	47.25	40.00	66.00	0.000006
PRWE after surgery	>60 years	38	9.50	1.00	27.00	

Abbreviations: N, group size; Q$_1$, lower quartile; Q$_3$, upper quartile; w, women; m, men.

Table 6. Results of the BCTQ analysis.

Scales	Group	N Valid	Median	Q$_1$	Q$_3$	*p*
SSS		77	1.09	1.00	1.64	0.000000
FSS		77	1.25	1.00	2.00	0.000001
SSS	w	66	1.18	1.00	1.64	0.000000
FSS	w	66	1.25	1.00	2.38	0.000005
SSS	m	11	1.00	1.00	1.45	0.008599
FSS	m	11	1.13	1.00	1.25	0.000352
SSS	<60 years	45	1.00	1.00	1.36	0.000000
FSS	<60 years	45	1.25	1.00	1.88	0.000000
SSS	>60 years	32	1.32	1.05	1.73	0.012840
FSS	>60 years	32	1.25	1.00	2.44	0.010862

Abbreviations: FSS, Functioning Status Scale; SSS, Symptom Severity Scale; w, women; m, men; N, group size; Q$_1$, lower quartile; Q$_3$, upper quartile. P-normality test for each group.

The mean score of the BCTQ for all subjects on the scale regarding the severity of symptoms was 1.48 ± 0.75 after surgery, and on the scale regarding function, it was 1.62 ± 0.85. There were no statistically significant differences in the scores between the male and female groups or between age groups ($p > 0.05$) (Table 6).

Chronic CTS patients' surgical results indicated that there was no correlation between the duration of symptoms and the questionnaire results (Table 7).

The correlation between the score of the questionnaires and the number of months between surgery and follow-up was also checked. Post-operative DASH, PRWE, and BCTQ scores improved with the number of months to wait for a follow-up visit.

Analysis of the questions about satisfaction after surgery in terms of symptom duration found that there was no significant interaction effect between these factors.

The time after which symptoms were retreated and patients' satisfaction was raised was 12 months on average.

Table 7. Results of the correlation analysis.

Correlation	N	r	p
Duration of symptoms & DASH	86	−0.01	0.951
Duration of symptoms & PRWE	86	−0.04	0.740
Duration of symptoms & BCTQ	77	0.05	0.662
Follow-up & DASH	86	**−0.26**	**0.014**
Follow-up & PRWE	86	**−0.29**	**0.006**
Follow-up & BCTQ	77	**−0.24**	**0.004**

Abbreviations: N, number of patients; r, Spearman's rank correlation coefficient; p, p value; Follow-up, number of months between surgery and follow-up. Bold letters indicate significant correlations.

4. Discussion

This study provided evidence on the positive results of CTS treatment in patients with delayed surgical intervention. Regarding improvements in daily activities and hand function postoperatively, we can conclude that 88% of patients were satisfied with the surgical outcomes. DASH scores decreased and PRWE questionnaire scores decreased significantly at $p < 0.001$. The same holds true for the BCTQ score regarding the severity of symptoms and improvement of the hand function after surgery. Functional studies revealed an increase in global grip strength postoperatively at $p < 0.001$, while no significant difference was detected in the measurements of sensory discrimination.

The purpose of this study was to answer the question of what results could be expected from delayed surgery in patients with chronic carpal tunnel syndrome, and what patients could expect from such surgery. The group consisted of people who had the condition for a minimum of 5 years (in a previous study, it was an average of 9 years). Many of these people had been treated conservatively, but despite the use of various methods (e.g., wrist orthoses, manual therapy, and injections), the complaints were unable to be resolved; thus, this study did not analyze those methods [27]. The average age of the patients was 58; therefore, the statistical analysis divided the group into those under and over 60. This age is also often the limit of professional activity, which may reflected the results in these two age groups.

Although the endoscopic method of carpal tunnel decompression has become increasingly popular in recent years, the open method is still just as effective. Publications from 2004 to 2020 have indicated that the endoscopic method reduces hospitalization time and pain around the scar [11,12,28]. Long Chen's study of 1596 patients indicated that both methods showed similar results in the resolution of symptoms, but patients treated with the endoscopic method regained better function in the hand faster and returned to work sooner [29].

There are several answers to the question of what kind of surgical results we can expect in patients with short surgical waiting times in the literature, which can be referred to when analyzing selected parameters. One such paper was published by Louie et al., which reported a study conducted on 211 patients with a minimum follow-up period of 10 years (from 11 to 17 years), confirming the prior finding that most patients are satisfied after open median nerve decompression, even in the long term after surgery [30].

Considering the data presented in Table 7, it can be concluded that it takes a long time for hand function to improve, and one year after surgery may sometimes be too short of a period to regain satisfactory improvement. However, it should be noted that although the correlation results were statistically significant, the correlation coefficients were very low; therefore, the above conclusions may not be sufficiently significant.

4.1. Patient's Satisfaction

One of the main goals of this study was to answer the question of whether patients would be satisfied after surgery, and whether the outcome of surgery would be satisfactory to them. Patients were mostly satisfied with the outcome of the surgery after receiving treatment for CCTS. They mentioned the elimination of night-time numbness and improved sensation and strength in the hand as the main reasons for satisfaction, among others. Those who were dissatisfied with the surgery also sometimes felt that there was an improvement in overall hand function or the elimination of night-time numbness. Patients cited a thick scar or tingling around the scar as reasons for dissatisfaction. In addition, in the literature, most patients were satisfied with the outcome of surgery, such as in the study by Louie mentioned above [30]. In the study by Michelotti et al., a numerical scale from 1 to 100 was used to assess patient satisfaction (the average score was 90.33). Such an assessment seems more detailed, but for a patient, the answer to a yes or no question may be more obvious. [31] Among the other issues addressed in the Japanese study were factors affecting satisfaction with surgery [32]. In the study, the authors showed that age and depression were factors that significantly affected satisfaction with the outcome of surgery. Other studies have also highlighted the significant impact of psychological factors on satisfaction, such as the work by Maempel, who studied 809 patients, of which 674 said they were satisfied with the surgery [32]. In his paper, he reported that although people in a worse mental state had lower levels of satisfaction, most such people were generally satisfied with the outcome of surgery.

4.2. Thenar Muscle Atrophy

Another parameter measured was the evaluation of the thenar muscles. One of the main muscles is the short thumb abductor muscle (Latin: Musculus abductor pollicis brevis: APB). It is located most superficially; therefore, in cases of prolonged compression of the median nerve, the APB may undergo atrophy, which, in the clinical picture, will result in emaciation of the thenar muscles. In our study, such atrophy was observed in 49% of patients with CCTS before surgery. However, studies on CCTS have clearly shown that atrophy of the APB muscle in particular is very common [33,34]. Moreover, because of this, many patients lose the ability to oppose their thumb, an activity which is essential for daily functioning. In one study, APB atrophy after surgery was found in 36% of the subjects. In such cases, the literature suggests performing surgery to regain thumb opposition, such as tendon transfer of the palmaris longus muscle (PL) using the Camitz technique [33,35] or modified Camitz technique [36]. In our study, we used a self-reported scale to assess thenar muscles atrophy, where 1 meant normal thenar muscles and 0 meant thenar muscle significantly denervated.

4.3. Hand Grip Strength

Another characteristic symptom reported by patients with CTS is weakened strength in the hand, which often results, for example, in patients dropping objects [37]. One reason for this weakness is leanness of the APB. In this study, a dynamometer was used to measure global grip strength. Weakened grip strength can affect up to 71% of people with CTS (according to a study on 172 patients) [38]. One limitation of this measurement is the large effect of flexor muscles on global grip strength.

In the present study, a significant improvement in grip strength was observed in all of the studied patients. This improvement was statistically significant in women; however, no such improvement was observed in the male group. Studies on the electrophysiological measurements of the APB have also indicated improved parameters after surgery [34]. Capasso's work on severe CTS demonstrated that the strength of the hand improved after surgery [34]. With a significant improvement in strength in the hand, post-operative patients can perform more activities not possible before surgery, which certainly affects their satisfaction with the operation. Lai and his team tested whether flexor retinaculum reconstruction can affect grip strength and hand function (measured using the BCTQ) [39]. After analyzing 615 patients, they noted that such modification had no effect on global grip strength or SSS scores relating to symptom severity. The only significant findings were improvements in scores on the FSS related to patient functioning. In a prospective study by Michelotti et al. on outcomes after open CTS surgery, no statistically significant improvements were seen in measures of strength [40]. In the study group of 30 subjects, the average strength before surgery was 4.52, whereas 2 years after surgery, it was 4.9.

4.4. Sensation Studies Results

Sensory two-point discrimination (2PD) was measured using a Weber discriminator; the values are given in millimeters. In some subjects, the result was >15 mm (indeterminate sensory resolution). In this group of subjects, 15 mm was taken as the result for the purpose of statistical calculations.

In this study, in the measurement of the mean discrimination on the fingers of the operated hand, no significant improvement was observed; sensation on the fingertips remained comparable to that before surgery. Several results in the literature also showed no statistically significant improvement in 2PD scores. A study by Bai et al. evaluating 85 patients before and after CTS surgery (follow-up: 12 months after surgery) reported the following results: 6.9 mm before surgery and 3.1 mm after surgery [41]. A study by Wessel et al. involving 73 patients examined before and after surgery, in follow-up examinations performed, on average, more than 12 months after surgery, also showed no significant improvement in the 2PD test [20]. Before surgery, the average was 7.4 mm, whereas after surgery, it was 6.4 mm.

Although sensory abnormalities on the fingers affect the vast majority of patients with CTS (70–87%), the above studies indicate that sensory resolution is not an indicator of the surgical outcome [38]. Although some patients showed an indeterminate resolution, they were able to define a single-point sensing, which we interpreted as protective sensation, which is often sufficient for daily functioning. Discrimination sensation, however, is less important than the protective sensation necessary for daily functioning.

4.5. Specific Symptoms

During the physical examination in medical practice, the Tinel's sign and Phalen's sign, due to their ease of performance, are among the tests most often performed during the diagnosis of CTS. These tests are not highly sensitive and specific; therefore, a positive result does not always mean a diagnosis of CTS.

In patients with CTS, the Tinel's sign has a lower sensitivity and specificity than the Phalen's sign. Hegmann's study, mentioned above, showed a sensitivity of Phalen's sign of 52.8% and Tinel's sign of 37.7% [42]. A tingling sensation in the hand waking patients from sleep was reported by 77.4% of people. In contrast, George Phalen's original study of 654 patients showed the sensitivity of Phalen's symptom to be 74% [43].

In this study, the majority of patients were positive for both tests before surgery, which indeed became negative after surgery.

4.6. Questionnaires

To assess hand function, patients completed three questionnaires. A decrease in the questionnaire score after surgery indicated an improvement in overall function. In the literature, the Boston Carpal Tunnel Questionnaire is most commonly used to assess the severity of symptoms and hand function [44]. The other tests are also suitable for reliably assessing patients with CTS, but the BCTQ is the shortest of them. Greenslade, in his study, compared the time it took patients to complete questionnaires assessing hand function [27]. The DASH questionnaire took an average of 6.8 min, whereas the BCTQ took 5.6 min. In this study, nonparametric tests were used to calculate the final results of the questionnaires; the medians and quartile ranges are reported in the tables. However, the results of the mean are discussed further because these values are most commonly reported in the literature. In this study, the mean score of the DASH questionnaire decreased from a preoperative score of 44.82 to 14.12 after surgery, which means that patients reported fewer symptoms after surgery and their upper limb function improved. Patients under 60 years of age had better results, meaning they regained better function than the older group of patients. The results of the questionnaire from the work and sports/playing an instrument module were not included in this paper due to the insufficient number of completed questionnaires. In the publication by Bai et al. [41] regarding 85 patients with CTS lasting approximately 6 months, the DASH score before surgery was 35.1, whereas after open median nerve decompression surgery, it was 10.4; thus, each group had better scores, i.e., greater dexterity in the hand, than the patients in our study. In the BCTQ, they also had better scores: before the operation, the mean score on the SSS was 2.7, and on the FSS, it was 2.4; after the operation, the mean score on the SSS was 1.3, and on the FSS, it was 1.1, although the improvement in scores on this questionnaire was not statistically significant.

In the results of the PRWE questionnaire, the whole group significantly improved after surgery, i.e., their wrist function increased. In addition, this questionnaire showed that after surgery, statistically better results were recorded in the group of patients under the age of 60 than in the older group. The only publication on patients with CTS who completed the PRWE questionnaire concerns conservative treatment [45]. In the literature, this questionnaire is mainly used to assess wrist function after fractures of the radius and elbow; thus, after reviewing the publications, its selection for the evaluation of patients with CTS may not be appropriate [46].

The BCTQ is a questionnaire that is dedicated to patients with CTS.

Our study was conducted in a group of patients after surgery. In Louie's study, the mean of the BCTQ (published as the Levine-Katz scale) was 1.3 on the SSS (13% of patients had scores $\geq$2 points, indicating severe symptoms) and 1.6 on the FFS scale (26% had scores $\geq$2 points, indicating poor function) [30].

In a study by Michelotti et al. conducted 2 years after surgery, the SSS score was 2.63 before surgery and 1.23 after surgery, whereas the FSS score was 2.24 and 1.16 before and after surgery, respectively [40].

5. Study Limitations

Notably, this study has some limitations which should be mentioned. The diagnosis of CTS was based on history and clinical symptoms. The purpose of this study was not to verify the diagnosis of CTS and analyze electroneurographic (ENG) or electromyographic (EMG) results, but to answer the question of how effective surgical treatment is in CCTS cases. We now also believe that it would be valuable to complete the BCTQ preoperatively as well.

We considered the functional hand grip test to be more reliable than manual muscle testing of the abductor pollicis brevis muscle. The patients in our study represented different degrees of thenar muscle atrophy; therefore, we decided on a simple evaluation of this symptom in a two-stage (0—present or 1—absent) score. Moreover, we used global grip strength evaluation, which is more common in the literature, but it is questioned as inferior to pinch grip measurements.

Another issue is the deliberate omission of an analysis of the conservative treatment applied. Such information on whether all patients were treated with physical methods, steroid injections, or wrist orthoses could have been an additional value of the paper, which would be especially of interest to physiotherapists. However, assuming that conservative methods were ineffective, these data were not analyzed in this study.

6. Conclusions

Patients with chronic CTS report better hand function after surgery despite the long duration of symptoms. Selected parameters (thenar muscles degeneration, global hand grip strength, and characteristic symptoms such as numbness or pain) were improved or retreated, but sensory resolution after surgery did not change and was similar to pre-surgery values.

The severity of symptoms and hand function after surgery improved relative to the pre-operative results, and the results of questionnaires assessing this function were satisfactory.

The surgical treatment of patients with chronic carpal tunnel syndrome provided results according to which the majority of respondents were satisfied, and the findings from the self-report questionnaire indicated a reduction in the discomfort experienced before surgery.

Author Contributions: Conceptualization, M.T. and L.R.; methodology, M.T., P.C. and L.R.; software, M.T., M.J. and E.B.; validation, M.T., P.C. and L.R.; formal analysis, M.T., P.C. and L.R.; investigation, M.T., M.J. and E.B.; resources, M.T., M.J. and E.B.; data curation, M.T., M.J. and E.B.; writing—original draft preparation, M.T. and P.C.; writing—review and editing, M.T., P.C. and J.H.; visualization, M.T. and J.H.; supervision, J.H. and L.R.; project administration, M.T.; funding acquisition, M.T. All authors have read and agreed to the published version of the manuscript.

Funding: This research received no external funding.

Institutional Review Board Statement: The study was approved by the Bioethics Committee at the Poznan University of Medical Sciences (Resolution no. 32/15, date of approval 8 January 2015).

Informed Consent Statement: Informed consent was obtained from all subjects involved in the study.

Data Availability Statement: Data available on request due to restrictions e.g., privacy or ethical.

Conflicts of Interest: The authors declare no conflict of interest.

References

1. Trumble, T.E.; Budoff, J.E.; Cornwall, R. (Eds.) *Core Knowledge in Orthopaedics: Hand, Elbow, and Shoulder;* Mosby, Inc.: Philadelphia, PA, USA, 2005.
2. Newington, L.; Harris, E.C.; Walker-Bone, K. Carpal tunnel syndrome and work. *Best Pract. Res. Clin. Rheumatol.* **2015**, *29*, 440–453. [CrossRef]
3. Lee, J.-K.; Lee, S.-H.; Kim, B.; Jung, K.; Park, I.; Han, S.-H. Risk Factors of Carpal Tunnel Syndrome for Male Patient Undergoing Carpal Tunnel Release. *Handchir. Mikrochir. Plast. Chir.* **2018**, *50*, 335–340. [CrossRef]
4. Aboonq, M.S. Pathophysiology of carpal tunnel syndrome. *Neurosci. J.* **2015**, *20*, 4–9.
5. MacDermid, J.; Wessel, J. Clinical diagnosis of carpal tunnel syndrome: A systematic review. *J. Hand Ther. Off. J. Am. Soc. Hand Ther.* **2004**, *17*, 309–319. [CrossRef] [PubMed]
6. Padua, L.; Coraci, D.; Erra, C.; Pazzaglia, C.; Paolasso, I.; Loreti, C.; Caliandro, P.; Hobson-Webb, L.D. Carpal tunnel syndrome: Clinical features, diagnosis, and management. *Lancet Neurol.* **2016**, *15*, 1273–1284. [CrossRef]
7. Phalen, G.S. The carpal-tunnel syndrome. Clinical evaluation of 598 hands. *Clin. Orthop. Relat. Res.* **1972**, *83*, 29–40. [CrossRef] [PubMed]
8. Petrover, D.; Richette, P. Treatment of carpal tunnel syndrome: From ultrasonography to ultrasound guided carpal tunnel release. *Jt. Bone Spine* **2018**, *85*, 545–552. [CrossRef]
9. Han, B.-S.; Kim, K.-H.; Kim, K.-J. The Radiographic Risk Factors Predicting Occurrence of Idiopathic Carpal Tunnel Syndrome in Simple Wrist X-ray. *J. Clin. Med.* **2022**, *11*, 4031. [CrossRef]
10. Sonoo, M.; Menkes, D.L.; Bland, J.D.P.; Burke, D. Nerve conduction studies and EMG in carpal tunnel syndrome: Do they add value? *Clin. Neurophysiol. Pract.* **2018**, *3*, 78–88. [CrossRef]
11. Zhao, H.; Zhao, Y.; Tian, Y.; Yang, B.; Qiu, G.-X. Comparison of endoscopic versus open surgical treatment of carpal tunnel syndrome. *Acta Acad. Med. Sin.* **2004**, *26*, 657–660.
12. Li, Y.; Luo, W.; Wu, G.; Cui, S.; Zhang, Z.; Gu, X. Open versus endoscopic carpal tunnel release: A systematic review and meta-analysis of randomized controlled trials. *BMC Musculoskelet. Disord.* **2020**, *21*, 272. [CrossRef] [PubMed]
13. De Smet, L.; Steenwerckx, A.; Van den Bogaert, G.; Cnudde, P.; Fabry, G. Value of clinical provocative tests in carpal tunnel syndrome. *Acta Orthop. Belg.* **1995**, *61*, 177–182. [CrossRef]
14. Richter, M.; Brüser, P. Value of clinical diagnosis in carpal tunnel syndrome. *Handchir. Mikrochir. Plast. Chir.* **1999**, *31*, 373–376, discussion 377. [CrossRef] [PubMed]
15. Pavesi, G.; Olivieri, M.F.; Misk, A.; Mancia, D. Clinical-electrophysiological correlations in the carpal tunnel syndrome. *Ital. J. Neurol. Sci.* **1986**, *7*, 93–96. [CrossRef] [PubMed]
16. Aghda, A.K.; Asheghan, M.; Amanollahi, A. Comparisons of electrophysiological and clinical findings between young and elderly patients with Carpal Tunnel Syndrome. *Rev. Neurol.* **2020**, *176*, 387–392. [CrossRef]
17. Elshebawy, H.; Ezzeldin, M.Y.; Elzamarany, E.H. Characteristics of COVID and post COVID polyneuropathies in adults and pediatrics: An Egyptian sample. *Egypt. J. Neurol. Psychiatry Neurosurg.* **2021**, *57*, 178. [CrossRef]
18. Córdova-Martínez, A.; Caballero-García, A.; Pérez-Valdecantos, D.; Roche, E.; Noriega-González, D.C. Peripheral Neuropathies Derived from COVID-19: New Perspectives for Treatment. *Biomedicines* **2022**, *10*, 1051. [CrossRef]

19. Saif, A.; Pick, A. Polyneuropathy following COVID-19 infection: The rehabilitation approach. *BMJ Case Rep.* **2021**, *14*, e242330. [CrossRef]
20. Wessel, L.E.; Ekstein, C.M.; Marshall, D.C.; Chen, A.Z.; Osei, D.A.; Fufa, D.T. Pre-operative Two-Point Discrimination Predicts Response to Carpal Tunnel Release. *HSS J.* **2020**, *16*, 206–211. [CrossRef]
21. Czarnecki, P.; Wawrzyniak-Bielęda, A.; Romanowski, L. Polish Adaptation of Wrist Evaluation Questionnaires. *Ortop. Traumatol. Rehabil.* **2015**, *17*, 241–248. [CrossRef]
22. Golicki, D.; Krzysiak, M.; Strzelczyk, P. Translation and cultural adaptation of the Polish version of the Disabilities of the Arm, Shoulder and Hand (DASH) and QuickDASH questionnaires. *Ortop. Traumatol. Rehabil.* **2014**, *16*, 387–395. [CrossRef] [PubMed]
23. Greenslade, J.R.; Mehta, R.L.; Belward, P.; Warwick, D.J. Dash and Boston questionnaire assessment of carpal tunnel syndrome outcome: What is the responsiveness of an outcome questionnaire? *J. Hand Surg. Br.* **2004**, *29*, 159–164. [CrossRef]
24. Mulders, M.A.; Kleipool, S.C.; Dingemans, S.A.; van Eerten, P.V.; Schepers, T.; Goslings, J.C.; Schep, N.W. Normative data for the Patient-Rated Wrist Evaluation questionnaire. *J. Hand Ther.* **2018**, *31*, 287–294. [CrossRef] [PubMed]
25. De Kleermaeker, F.G.C.M.; Boogaarts, H.D.; Meulstee, J.; Verhagen, W.I.M. Minimal clinically important difference for the Boston Carpal Tunnel Questionnaire: New insights and review of literature. *J. Hand Surg. Eur. Vol.* **2018**, *44*, 283–289. [CrossRef]
26. Leite, J.C.; Jerosch-Herold, C.; Song, F. A systematic review of the psychometric properties of the Boston Carpal Tunnel Questionnaire. *BMC Musculoskelet. Disord.* **2006**, *7*, 78. [CrossRef]
27. Pintucci, M.; Imamura, M.; Thibaut, A.; De Exelnunes, L.M.; Minagato, M.M.; Kaziyama, H.H.; Mamura, S.T.; Stecco, A.; Fregni, F.; Battistella, L.R. Evaluation of fascial manipulation in carpal tunnel syndrome: A pilot randomized clinical trial. *Eur. J. Phys. Rehabil. Med.* **2017**, *53*, 630–631. [CrossRef] [PubMed]
28. Vasiliadis, H.S.; Nikolakopoulou, A.; Shrier, I.; Lunn, M.P.; Brassington, R.; Scholten, R.J.; Salanti, G. Endoscopic and Open Release Similarly Safe for the Treatment of Carpal Tunnel Syndrome. A Systematic Review and Meta-Analysis. *PLoS ONE* **2015**, *10*, e0143683. [CrossRef]
29. Chen, L.; Duan, X.; Huang, X.; Lv, J.; Peng, K.; Xiang, Z. Effectiveness and safety of endoscopic versus open carpal tunnel decompression. *Arch. Orthop. Trauma. Surg.* **2014**, *134*, 585–593. [CrossRef]
30. Louie, D.L.; Earp, B.E.; Collins, J.E.; Losina, E.; Katz, J.N.; Black, E.M.; Simmons, B.P.; Blazar, P.E. Outcomes of open carpal tunnel release at a minimum of ten years. *J. Bone Jt. Surg. Am.* **2013**, *95*, 1067–1073. [CrossRef]
31. Michelotti, B.M.; Vakharia, K.T.; Romanowsky, D.; Hauck, R.M. A Prospective, Randomized Trial Comparing Open and Endoscopic Carpal Tunnel Release Within the Same Patient. *Hand* **2020**, *15*, 322–326. [CrossRef]
32. Bae, J.-Y.; Kim, J.K.; Yoon, J.O.; Kim, J.H.; Ho, B.C. Preoperative predictors of patient satisfaction after carpal tunnel release. *Orthop. Traumatol. Surg. Res.* **2018**, *104*, 907–909. [CrossRef]
33. Maempel, J.F.; Jenkins, P.J.; McEachan, J.E. The relationship of mental health status to functional outcome and satisfaction after carpal tunnel release. *J. Hand. Surg. Eur. Vol.* **2020**, *45*, 147–152. [CrossRef] [PubMed]
34. Nanno, M.; Kodera, N.; Tomori, Y.; Takai, S. Minimally invasive modified Camitz opponensplasty for severe carpal tunnel syndrome. *J. Orthop. Surg.* **2018**, *26*, 2309499018770914. [CrossRef]
35. Capasso, M.; Manzoli, C.; Uncini, A. Management of extreme carpal tunnel syndrome: Evidence from a long-term follow-up study. *Muscle Nerve* **2009**, *40*, 86–93. [CrossRef]
36. Foucher, G.; Malizos, C.; Sammut, D.; Braun, F.M.; Michon, J. Primary palmaris longus transfer as an opponensplasty in carpal tunnel release. A series of 73 cases. *J. Hand Surg. Br.* **1991**, *16*, 56–60. [CrossRef] [PubMed]
37. Park, I.-J.; Kim, H.-M.; Lee, S.-U.; Lee, J.-Y.; Jeong, C. Opponensplasty using palmaris longus tendon and flexor retinaculum pulley in patients with severe carpal tunnel syndrome. *Arch. Orthop. Trauma Surg.* **2010**, *130*, 829–834. [CrossRef]
38. Günther, C.M.; Bürger, A.; Rickert, M.; Crispin, A.; Schulz, C.U. Grip strength in healthy caucasian adults: Reference values. *J. Hand Surg. Am.* **2008**, *33*, 558–565. [CrossRef]
39. Banach, M.; Gribbin, D.; Zajaczkowska, A. Clinical and electrophysiological correlations concerning patients with carpal tunnel syndrome. *Przegl. Lek.* **2010**, *67*, 692–696.
40. Lai, S.; Zhang, K.; Li, J.; Fu, W. Carpal tunnel release with versus without flexor retinaculum reconstruction for carpal tunnel syndrome at short- and long-term follow up-A meta-analysis of randomized controlled trials. *PLoS ONE* **2019**, *14*, e0211369. [CrossRef] [PubMed]
41. Bai, J.; Kong, L.; Zhao, H.; Yu, K.; Zhang, B.; Zhang, J.; Tian, D. Carpal tunnel release with a new mini-incision approach versus a conventional approach, a retrospective cohort study. *Int. J. Surg.* **2018**, *52*, 105–109. [CrossRef]
42. Hegmann, K.T.; Merryweather, A.; Thiese, M.S.; Kendall, R.; Garg, A.; Kapellusch, J.; Foster, J.; Drury, D.; Wood, E.M.; Melhorn, J.M. Median Nerve Symptoms, Signs, and Electrodiagnostic Abnormalities Among Working Adults. *J. Am. Acad. Orthop. Surg.* **2018**, *26*, 576–584. [CrossRef] [PubMed]
43. Phalen, G.S. The carpal-tunnel syndrome. Seventeen years' experience in diagnosis and treatment of six hundred fifty-four hands. *J. Bone Jt. Surg. Am.* **1966**, *48*, 211–228. [CrossRef]
44. De Kleermaeker, F.G.C.M.; Meulstee, J.; Claes, F.; Bartels, R.H.M.A.; Verhagen, W.I.M. Outcome after carpal tunnel release: Effects of learning curve. *Neurol. Sci.* **2019**, *40*, 1813–1819. [CrossRef]

45. Chen, L.; Xue, L.; Li, S.; Kang, T.; Chen, H.; Hou, C. Clinical research on mild and moderate carpal tunnel syndrome treated with contralateral needling technique at distal acupoints and acupuncture at local acupoints. *Chin. Acupunct. Moxibustion* **2017**, *37*, 479–482. [CrossRef]
46. Changulani, M.; Okonkwo, U.; Keswani, T.; Kalairajah, Y. Outcome evaluation measures for wrist and hand: Which one to choose? *Int. Orthop.* **2008**, *32*, 1–6. [CrossRef]

Article

Influence of Risk Factors on the Well-Being of Elderly Women with Knee Osteoarthritis

Ivana Minaković [1,2,*], Jelena Zvekić Svorcan [1,3], Tanja Janković [1,3], Hajdana Glomazić [4], Mirjana Smuđa [1,5], Dejan Živanović [1,6], Jovan Javorac [1,7] and Bela Kolarš [1,2]

[1] Faculty of Medicine, University of Novi Sad, 21000 Novi Sad, Serbia; jelena.zvekic-svorcan@mf.uns.ac.rs (J.Z.S.); tanja.jankovic@mf.uns.ac.rs (T.J.); mirjana.smudja@uns.ac.rs (M.S.); dejan.zivanovic@mf.uns.ac.rs (D.Ž.); jovan.javorac@mf.uns.ac.rs (J.J.); 902009d22@mf.uns.ac.rs (B.K.)

[2] Health Center "Novi Sad", 21000 Novi Sad, Serbia

[3] Special Hospital for Rheumatic Diseases Novi Sad, 21000 Novi Sad, Serbia

[4] Institute of Criminological and Sociological Research, 11000 Belgrade, Serbia; hajdana.ng@gmail.com

[5] Department of Higher Medical School, The Academy of Applied Studies Belgrade, 11000 Belgrade, Serbia

[6] Department of Psychology, College of Social Work, 11000 Belgrade, Serbia

[7] Institute for Pulmonary Diseases of Vojvodina, 21000 Novi Sad, Serbia

* Correspondence: ivana.minakovic@mf.uns.ac.rs

Citation: Minaković, I.; Svorcan, J.Z.; Janković, T.; Glomazić, H.; Smuđa, M.; Živanović, D.; Javorac, J.; Kolarš, B. Influence of Risk Factors on the Well-Being of Elderly Women with Knee Osteoarthritis. *Medicina* **2023**, *59*, 1396. https://doi.org/10.3390/medicina59081396

Academic Editors: Dejan Nikolic and Milica Lazovic

Received: 28 June 2023
Revised: 23 July 2023
Accepted: 25 July 2023
Published: 29 July 2023

Abstract: *Background and Objectives:* Knee osteoarthritis (KOA) is a widespread chronic joint disease characterized by functional limitations and pain. Functioning restrictions exert a detrimental impact on societal integration, relationships, and psychological well-being, resulting in significant emotional distress in KOA patients. The objective of this study is to examine how various risk factors impact the emotional well-being of individuals with KOA. *Materials and Methods:* This prospective cross-sectional study involved 154 postmenopausal women treated at the Special Hospital for Rheumatic Diseases in Novi Sad, Serbia. The experimental group comprised 97 individuals with chronic knee pain and structural knee damage (Kellgren–Lawrence (KL) scale II-IV), while the control group had 53 individuals with chronic knee pain but no structural knee damage (KL scale 0-I). The collected data consisted of sociodemographic factors, general characteristics, associated diseases, and laboratory results. Adequate anthropometric measurements were conducted, and all subjects were required to complete the SF-36 RAND questionnaire. *Results:* The analysis identified several variables that independently influenced emotional well-being. These included pain intensity (beta (β) 0.21; 95% CI: 0.03–0.20; $p < 0.01$), social functioning (beta (β) 0.47; 95% CI: 0.23–0.43; $p < 0.001$), physical functioning (beta (β) 0.23; 95% CI: 0.04–0.21; $p < 0.01$), and education level (8–12 years: beta (β) 0.25; 95% CI: 1.47–9.41; $p < 0.01$; >12 years: beta (β) 0.27; 95% CI: 2.51–12.67; $p < 0.01$). However, the multivariate model revealed that only social functioning (beta (β) 0.57; 95% CI: 0.27–0.53; $p < 0.001$) and education level (8–12 years: beta (β) 0.21; 95% CI: 1.10–8.260; $p < 0.05$; >12 years: beta (β) 0.21; 95% CI: 1.18–10.30; $p < 0.05$) were significantly associated with emotional well-being in KOA patients. *Conclusions:* The findings of this study indicate that a reduced social functioning and a lower educational attainment are linked to a poorer emotional well-being among patients with KOA.

Keywords: osteoarthritis; knee; predictors; social functioning; pain; physical function; mental health

1. Introduction

Osteoarthritis (OA) is a diverse condition with various phenotypes [1]. It can affect any joint in the human body, although it primarily affects peripheral joints such as the knee, hip, hand, and spine [2,3]. The global prevalence of OA is approximately 350 million people [4], and it is commonly observed in one out of every three individuals over the age of 65 [5,6]. This disease is more prevalent among women, and obese women face a significant lifetime risk of up to 23.8% for symptomatic KOA [1]. Given its widespread occurrence, OA not only affects the individual's health, but also has a substantial impact on the overall population's health [1,7].

The presence of structural and functional limitations in patients with OA significantly contributes to pain and disability [8–10]. These symptoms are widely recognized as key manifestations of the disease [9,11]. Moreover, individuals with OA experience a considerable decline in their quality of life (QOL) due to diminished social connections, relationships, and emotional well-being [8]. Consequently, there is an increased risk of developing mental disorders [10,12]. These findings are consistent with a multitude of epidemiological studies, indicating a robust interconnection between physical diseases and common mental health conditions. They exhibit a high co-morbidity, meaning they often co-occur in individuals, and share crucial pathways that contribute to the development of ill health and disease [13].

In general, health is commonly perceived as the absence of illness or disease. These notions have been embedded in healthcare models, which tend to focus only on alleviating impairment and distress [13]. Presently, there are various treatment choices for KOA, including behavioral modifications like weight loss, exercise, and physical therapy coupled with pain-relieving medications, as well as surgical interventions [9]. Although numerous options are accessible, certain patients may still encounter insufficient symptom management, while others might experience adverse effects from the available interventions [14]. Among the surgical options, orthopedic surgery with total arthroplasty is recognized as a highly effective approach for addressing KOA [15]. However, it is essential to note that the surgery still has a failure rate that can affect the patient's psychology [16].

The impact of OA on mental health is particularly evident in the heightened vulnerability to conditions such as depression [1,12] and anxiety disorders [17]. This aspect is of the utmost significance since poor mental health, particularly the presence of depression, has been linked to unfavorable treatment outcomes in OA patients [12,18], as well as increasing the perception of pain [9]. Given that well-being encompasses various dimensions, including mental, physical, social, and environmental aspects [17,19], it becomes evident that placing a greater emphasis on providing psychological support to OA patients is crucial not only for pain reduction and functional improvement, but also for overall health enhancement.

Nevertheless, there is a scarcity of research regarding the factors that predict emotional well-being in patients with KOA. The aim of this study was to identify these predictors, thereby allowing for personalized patient treatment tailored to their specific characteristics, with a focus on improving modifiable risk factors.

2. Materials and Methods

2.1. Participants

A prospective cross-sectional study involved 154 postmenopausal women who were receiving treatment for chronic knee pain at the Special Hospital for Rheumatic Diseases Novi Sad, Serbia, between February 2022 and March 2023. The study was conducted in accordance with the guidelines of the Declaration of Helsinki. The study protocol obtained approval from the Ethics Committee of the Special Hospital for Rheumatic Diseases Novi Sad, Serbia (approval number: 14/29–3/1–21) and the Ethics Committee of the Faculty of Medicine Novi Sad, Serbia (approval number: 01–39/109/1). The informed consent was obtained from all participants.

Inclusion criteria: Women in postmenopause, aged 60–75 years who reported knee pain of intensity ≥ 3 on the numeric rating scale (NRS) persisting for a minimum duration of 3 months. Exclusion criteria: Participants with a history of knee surgery or injury, as well as presence of inflammatory rheumatic disease, metabolic joint diseases (such as gout and chondrocalcinosis), neuromuscular diseases (such as muscular dystrophies and myasthenia gravis) and participants who are currently being treated for a malignant disease. Furthermore, patients undergoing physical therapy, receiving corticosteroid therapy, or undergoing intraarticular chondroprotective therapy within the past three and six months, respectively, were also excluded.

At the outset, the study comprised 154 subjects, but four of them were later excluded upon the diagnosis of inflammatory rheumatic disease.

All patients suffering from chronic knee pain, who were referred to the Special Hospital for Rheumatic Diseases in Novi Sad, underwent an initial examination conducted by a physical medicine specialist. In the first phase of the study, sociodemographic information and data concerning related medical conditions were gathered, and all participants completed the SF-36-RAND questionnaire [20]. Body mass and height were measured using a precise digital weight scale and an adult height scale, respectively. Obesity level was determined by calculating the body mass index (BMI), which was obtained by dividing weight in kilograms by height in square meters. Waist circumference was measured with a measuring tape, and knee range of motion (ROM) was assessed using a goniometer. A knee flexion $\geq$ 110 and an extension of 0 were considered as satisfactory ROM [21]. Blood pressure was assessed while the individual was in a seated and relaxed position, using a validated and calibrated Riester sphygmomanometer on both hands. After a short interval, the measurement was repeated on the hand that initially showed the higher pressure reading [22]. In the second phase of the study, all participants underwent bilateral anteroposterior and lateral X-rays of both knees. A skilled radiologist assessed the degree of radiological damage using both the Kellgren–Lawrence (KL) scale and the Altman atlas. Subsequently, the subjects were classified into two groups based on these evaluations. The experimental group consisted of 97 subjects who showed radiological damage classified as grade II-IV on the KL scale, while the control group included 53 patients who had no radiological changes in their knees (KL scale 0-I).

In the third phase of the study, fasting glucose levels and lipid panel were measured to assess metabolic parameters. The state reference laboratory at the Special Hospital for Rheumatic Diseases, Novi Sad, Serbia, was responsible for conducting all of the laboratory analyses. Metabolic syndrome (MetS) was identified based on the criteria established by the International Diabetes Federation (IDF) Consensus statements [23]. MetS was defined as the combination of central obesity (waist circumference of 80 cm or more) and at least two of the following four factors: elevated triglyceride levels (equal to or greater than 150 mg/dL or 1.7 mmol/L) or receiving specific treatment for this lipid abnormality; reduced levels of HDL cholesterol (less than 50 mg/dL or 1.29 mmol/L in females) or receiving specific treatment for this lipid abnormality; systolic blood pressure equal to or greater than 130 mm Hg or diastolic blood pressure equal to or greater than 85 mm Hg or undergoing treatment for previously diagnosed hypertension; and elevated fasting plasma glucose levels (equal to or greater than 100 mg/dL or 5.6 mmol/L) or previously diagnosed type 2 diabetes mellitus (T2DM).

In the extension, a flow diagram is presented, illustrating the cross-sectional design used for data collection, as well as the inclusion and exclusion criteria (Figure 1).

2.2. Assessment of Emotional Well-Being, Social Functioning, Physical Functioning, and Pain Level

The SF-36, RAND questionnaire comprises 8 subscales encompassing physical function, limitations in daily activities due to physical and emotional health, energy levels and fatigue, emotional well-being, social functioning, pain, and general health. Each subscale is scored on a scale of 0 to 100, with higher scores indicating greater levels of well-being and better overall health [20].

2.3. Evaluation of the Educational Attainment Level

Participants were divided into three groups according to their educational attainment level. The first group consisted of female patients with ≤8 years of schooling (primary education in Serbia lasts 8 years). The second group consisted of female patients with 8–12 years of schooling (secondary education in Serbia lasts 4 years). The third group consisted of respondents with more than 12 years of education (completed at least higher-education schooling).

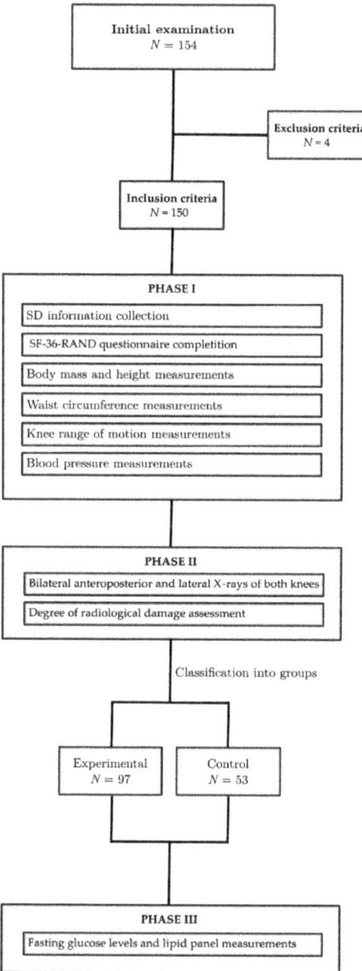

Figure 1. Cross-sectional design of the study.

2.4. Statistical Analysis

Considering the objectives of the study, the sample size was calculated with a confidence interval of 85% (consequently the power of the study was 85%), with a maximum error of 10% and a critical incidence value of 50%. As a result, a minimum of 52 respondents per group is required to meet these criteria. The statistical analysis involved examining differences among respondents using chi-squared tests and Mann–Whitney tests. All tests were conducted with a significance level of $p < 0.05$. To determine the relationship between the outcome variable (emotional well-being) and explanatory variables, a set of linear regression models (beta and 95% confidence interval—CI) was employed, considering both univariate and multivariate associations. The range of Cronbach's alpha, which was between 0.796 and 0.885, indicates good internal consistency. All analyses were performed with IBM SPSS ver. 25 (IBM Corp., Armonk, NY, USA).

3. Results

The study included 150 women with an average age of 67 (IQR = 8.0). Women with KOA exhibited a higher body mass index (BMI) of 30.4 (IQR = 6.0) compared to the control group with a BMI of 28.6 (IQR = 6.6), with a statistically significant difference ($p = 0.044$). Furthermore, MetS was found to be more prevalent in the experimental group (81.4%) compared to the control group (58.5%), with a significant difference ($p = 0.002$). Among the participants, 23.3% had T2DM, and 75.3% had hypertension. The number of comorbidities among KOA participants ranged from 0 to 6, with an average of 3 (IQR = 1.0) (Table 1).

Table 1. General characteristics of the respondents and frequent comorbidities.

	(ALL) N = 150	Experimental Group N = 97	Control Group N = 53	*p*-Value
Age (years), Me ± IQR (Min–Max)	67.0 ± 8.0 (60–75)	67.0 ± 7.0 (60–75)	65.0 ± 10.5 (60–75)	0.063 [b]
BMI (kg/m²), Me ± IQR (Min–Max)	30.0 ± 6.4 (18.5- 47.3)	30.4 ± 6.0 (19.8–47.3)	28.6 ± 6.6 (18.5–42.1)	0.044 [b]
Mets:				0.002 [a]
Yes	110 (73.3%)	79 (81.4%)	31 (58.5%)	
No	40 (26.7%)	18 (18.6%)	22 (41.5%)	
T2DM:				0.882 [a]
Yes	35 (23.3%)	23 (23.7%)	12 (22.6%)	
No	115 (76.7%)	74 (76.3%)	41 (77.4%)	
Hypertension:				0.246 [a]
Yes	113 (75.3%)	76 (78.4%)	37 (69.8%)	
No	37 (24.7%)	21 (21.6%)	16 (30.2%)	
The number of comorbidities, Me ± IQR (Min–Max)	3.0 ± 1.0 (0–7)	3.0 ± 1.0 (0–6)	2.0 ± 1.0 (1–7)	0.696 [b]

Note. [a] Chi-squared test. [b] Mann–Whitney.

The study determined the reliability of the pain, social functioning, and emotional well-being subscales to be 0.796, 0.827, and 0.885, respectively. The average score obtained by the participants on the pain subscale was 47.8 (standard deviation = 19.6), on the social functioning subscale was 68.2 (standard deviation = 15.5), and on the emotional well-being subscale was 62.2 (standard deviation = 10.8). The descriptive statistics for the individual subscales of the SF-36 questionnaire are presented in Table 2.

Table 2. Descriptive statistics of SF-36 scale.

SF-36 Domains	Me	IQR	M	SD	Skewness	Kurtosis	K-S	α
Pain	45.0	32.5	47.8	19.6	0.057	−0.765	0.156 **	0.796
Social functioning	75.0	15.6	68.2	15.5	−0.477	0.066	0.265 **	0.827
Emotional well-being	64.0	12.0	62.2	10.8	−0.839	1.092	0.130 **	0.885

Note. ** $p < 0.01$. Me = Median. IQR = Inter-quartile range. M = Mean. SD = Std. Deviation. K-S = Kolmogorov–Smirnov test. α = Cronbach's alpha.

Table 3 displays the univariate linear regression model involving the SF-36 RAND subscale of emotional well-being as the dependent variable. The impact of individual independent variables was examined, followed by their combined impact. The variables that individually explain emotional well-being are as follows: pain intensity (beta (β) = 0.21; 95% CI: 0.03–0.20; $p < 0.01$), social functioning (beta (β) = 0.47; 95% CI: 0.23–0.43; $p < 0.001$), physical functioning (beta (β) = 0.23; 95% CI: 0.04–0.21; $p < 0.01$), and education level (8–12 years: beta (β) = 0.25; 95% CI: 1.47–9.41; $p < 0.01$; >12 years: beta (β) = 0.27; 95% CI: 2.51–12.67; $p < 0.01$). Therefore, individuals with a better social functioning, improved physical functioning, and reduced pain tend to have better emotional well-being. Furthermore, respondents who have completed secondary school or higher education exhibit superior emotional well-being compared to those with primary education or less.

Table 3. General characteristics and health status as determinants of emotional well-being, univariate linear regression.

	Univariate Linear Regression
Osteoarthritis (ref.: without osteoarthritis)	
Yes	−0.06 (−5.12–2.20)
Number of comorbidities (continuous)	−0.37 (−3.84––1.37)
Range of motion (ref.: unsatisfactory functionality)	
Satisfactory functionality	−0.17 (−7.32––0.37)
KL scale (continuous)	−0.10 (−2.42–0.55)
Duration of menopause (continuous)	−0.09 (−0.45–0.12)
SF36_pain (continuous)	0.21 (0.03–0.20) **
SF36_ social functioning (continuous)	0.47 (0.23–0.43) ***
SF36_ physical functioning (continuous)	0.23 (0.04–0.21) **
Age, years (continuous)	−0.02 (−0.44–0.32)
Lives alone (ref.: no)	
Yes	0.08 (−2.00–6.29)
Education level (ref.: ≤8)	
8–12	0.25 (1.47–9.41) **
>12	0.27 (2.51–12.67) **

Note. Values represent beta coefficients with corresponding 95% confidence intervals. ** $p < 0.01$; *** $p < 0.001$.

The multivariate model, which incorporates only the variables that demonstrated a statistically significant impact on the dependent variable in the univariate model, reveals that certain variables are associated with emotional well-being. Specifically, social functioning (beta (β) = 0.57; 95% CI: 0.27–0.53; $p < 0.001$) and education level (8–12 years: beta (β) = 0.21; 95% CI: 1.10–8.260; $p < 0.05$; >12 years: beta (β) = 0.21; 95% CI: 1.18–10.30; $p < 0.05$) demonstrate a connection with an enhanced emotional well-being (presented in Table 4). In summary, a better social functioning and a higher level of education are associated with improved emotional well-being.

Table 4. The variables that individually explain emotional well-being, multivariate linear regression.

	Multivariate Linear Regression
SF36_pain (continuous)	−0.21 (−0.25–0.06)
SF36_social functioning (continuous)	0.57 (0.27–0.53) **
SF36_physical functioning (continuous)	0.13 (−0.13–0.14)
Education level (ref.: ≤8)	
8–12	0.21 (1.10–8.260) *
>12	0.21 (1.18–10.30) *

Note. Values represent beta coefficients with corresponding 95% confidence intervals. * $p < 0.05$; ** $p < 0.001$.

4. Discussion

Our study indicates that pain intensity, social functioning, physical functioning, and education level are significant individual predictors of emotional well-being among KOA patients. However, the multivariate analysis reveals that among postmenopausal women with KOA, only social functioning and education level emerge as significant predictors of emotional well-being.

Advanced age, obesity, and physical inactivity are recognized as risk factors for various chronic diseases that are prevalent among individuals with KOA. Previous estimates indicate that a significant proportion of patients with OA, ranging from 59% to 87%, have been diagnosed with at least one additional chronic disease. Furthermore, as many as 31% of individuals with OA have the burden of five or more comorbidities [5,24]. Patients with OA typically experience an average of 2.6 comorbidities of moderate to severe intensity [5,24], which was consistent with the findings of our study (3 (IQR = 1.0)). Besides experiencing knee problems, individuals with KOA often encounter other chronic conditions, predominantly cardiovascular and pulmonary issues, as well as hypertension and diabetes [25].

Among the participants in our study, hypertension and T2DM were the prevailing ailments, and their occurrences are demonstrated in Table 1. The experimental and control groups were comparable in terms of the prevalence of hypertension and T2DM and there was no statistically significant distinction in the number of comorbidities between the observed groups ($p = 0.696$). Nonetheless, there was a significant difference in the number of subjects with MetS in the experimental group. The clinical significance of comorbidities in OA is acknowledged due to their potential impact on clinical practices, outcomes, treatment strategies, prognosis, health-related quality of life (HRQOL), and healthcare costs [25,26]. Calders et al. documented that an increased prevalence of additional health conditions exacerbates both pain and functional limitations among individuals suffering from hip and knee OA [11]. Contrary to our expectations, we were unable to establish a statistically significant correlation between the number of comorbidities and emotional well-being in patients with KOA. This discrepancy may be attributed to variations in the severity and duration of specific diseases, i.e., symptoms of the disease. In our previous pilot study [27], the number of comorbidities emerged as a significant predictor of emotional well-being, but the limited sample size could have introduced bias into the results.

As anticipated, the majority of participants in our study exhibited an elevated body mass. In a comprehensive meta-analysis encompassing 131 studies, it was determined that the average body mass index (BMI) among individuals with OA was 28.2 [9]. We observed an even higher BMI (30.4 (IQR = 6.0)) among subjects with radiographic KOA. This finding is not surprising, considering that our study specifically focused on participants with KOA and it is widely recognized that being overweight or obese significantly increases the risk of developing OA, particularly in the knee joint [28].

In the majority of studies, advancing age is linked to a decline in QOL [8]. It has been observed that younger individuals have greater expectations when it comes to their QOL, likely due to their higher efficiency in performing daily activities, work, and leisure [8,29]. We were unable to establish a correlation between emotional well-being and the age of the respondents. This can be attributed to the homogeneity of the group in terms of age and gender.

Utilizing a univariate linear regression model, we discovered that pain, social functioning, physical functioning, and education level serve as predictors for emotional well-being among patients with KOA. This aligns with the findings of Giesbrecht et al., who demonstrated a connection between psychological well-being and the ability to perform daily activities [30]. Additionally, Nazarinasab et al. reported that the extent of disability is associated with mental health [12]. Unlike knee ROM, physical functioning (evaluated using the SF-36-RAND questionnaire) emerged as a statistically significant predictor of emotional well-being. It has been suggested that a range of 100–110 degrees in knee flexion is essential for everyday activities [21], and we adopted these values as a reference. Nevertheless, there are several other activities, such as entering a bathtub, engaging in gardening, performing DIY tasks, or changing socks, that require a greater knee ROM [21,31].

In accordance with the outcomes of our study, Fonseca-Rodrigues et al. have demonstrated that the intensity of pain is linked to the extent of emotional distress experienced by patients with OA [9], while Laires et al., reported that levels of pain often correlate with reduced levels of satisfaction with treatment [32]. The existing literature suggests that mental health levels are inversely associated with knee pain, regardless of any observable changes in X-ray images [12]. Consistent with these findings, our study revealed that the severity of radiological knee damage does not predict emotional well-being among patients with KOA.

The educational attainment of patients with KOA significantly influences their QOL. It is widely acknowledged that there is a connection between education level and KOA, as individuals with lower education often engage in repetitive work tasks that heighten the risk of developing the condition. Moreover, having a lower educational level is associated with a decline in self-perceived QOL [8]. In a similar vein, our study's findings indicate that respondents with a secondary school education or higher exhibit better emotional well-being when compared to those with only primary education or lower.

The correlation between social isolation and both mental and physical well-being has been previously emphasized [33]. Additionally, the importance of friends, family, and caregivers as vital providers of social support has been acknowledged, as they play a key role in helping individuals with OA maintain their independence and receive encouragement to participate in meaningful activities [34]. In our study, both univariate and multivariate regression models revealed that social isolation significantly predicts emotional well-being among individuals with KOA. It is well-established that pain and a decreased functionality can contribute to the risk of social isolation [33,35]. Moreover, individuals with OA face additional factors that increase the likelihood of social isolation, including anxiety, depression, kinesiophobia, physical inactivity, and a diminished self-efficacy, all of which impede their functional independence [33,36]. Additionally, the literature indicates that a considerable proportion of individuals aged 65 and older experience challenges related to low self-efficacy, depression, and a reduced sense of QOL and well-being, all of which compromise their personal independence [17,37].

Taking into account the higher susceptibility of women to developing knee and hand OA, particularly during the postmenopausal phase [9], we investigated the impact of postmenopausal duration on emotional well-being. However, this specific variable did not emerge as a statistically significant risk factor.

Positive emotions and overall well-being play a significant role in affecting the occurrence of various health conditions and diseases, as well as the risk of premature mortality [13]. Previously, there have been suggestions that well-being might influence physiological processes that are pertinent to the risk of developing arthritis or the manifestation of its symptoms [38]. Considering the association of OA with detrimental effects on healthcare utilization and escalating personal and societal expenses, [34] it becomes highly imperative to conduct an examination and understanding of the contributing factors that lead to a reduced emotional well-being. We believe that our research will provide valuable insights to enhance the management of KOA.

This study possesses some limitations. Firstly, the patients included in the study were not representative of the entire population of our country, but rather comprised patients from Vojvodina, a Serbian province, who were treated at a single healthcare facility. Secondly, important data on socioeconomic status, which is a significant factor and mediator for emotional well-being, were not collected during the enrollment process. However, this study was conducted prospectively, ensuring that every patient received a knee x-ray, and the extent of radiological damage was evaluated by a skilled radiologist using the KL scale and the Altman atlas. The data were collected by one trained researcher under consistent conditions throughout the study.

5. Conclusions

Our study revealed that poorer social functioning and lower educational attainment are associated with a higher risk of experiencing negative emotional well-being among patients with KOA. Typically, the treatment approach for KOA involves lifestyle modifications such as weight loss, exercise, physical therapy, pain-relieving medications, and, in some cases, surgical intervention. However, it is noteworthy that only a small proportion (estimated at 12%) of these interventions address the crucial aspect of social integration and support [38]. Therefore, it is imperative to emphasize the implementation of various strategies for social engagement, community participation, and specific forms of psychological support.

Author Contributions: Conceptualization, I.M., B.K. and J.Z.S.; methodology, I.M. and J.Z.S.; software, H.G. and B.K.; formal analysis, H.G, D.Ž. and J.J.; investigation, I.M; data curation, I.M., J.Z.S. and M.S.; writing—original draft preparation, I.M. and T.J.; writing—review and editing, D.Ž. and M.S.; visualization, T.J and J.J.; supervision J.Z.S. and T.J. All authors have read and agreed to the published version of the manuscript.

Funding: This research received no external funding.

Institutional Review Board Statement: The study was conducted in accordance with the Declaration of Helsinki and approved by the Ethics Committee of the Special Hospital for Rheumatic Diseases Novi Sad, Serbia (approval number: 14/29–3/1–21) and the Ethics Committee of the Faculty of Medicine Novi Sad, Serbia (approval number: 01–39/109/1), date of approval 14 September 2021 and 29 October 2021, respectively.

Informed Consent Statement: Informed consent was obtained from all subjects involved in the study.

Data Availability Statement: Not applicable.

Conflicts of Interest: The authors declare no conflict of interest.

References

1. Bortoluzzi, A.; Furini, F.; Scirè, C.A. Osteoarthritis and its management - Epidemiology, nutritional aspects and environmental factors. *Autoimmun. Rev.* **2018**, *17*, 1097–1104. [CrossRef] [PubMed]
2. Zvekić Svorcan, J.; Minaković, I.; Vojnović, M.; Miljković, A.; Mikov, J.; Bošković, K. The role of metabolic syndrome in the development of osteoarthritis. *Med. Pregl.* **2022**, *75*, 39–43. [CrossRef]
3. Zvekić Svorcan, J.; Minaković, I.; Krasnik, R.; Mikić, D.; Mikov, J.; Stamenković, B. Structural damage of the hand in hand osteoarthritis: Impact on function, pain, and satisfaction. *Hippokratia* **2022**, *26*, 7–12. [PubMed]
4. Callahan, L.F.; Cleveland, R.J.; Allen, K.D.; Golightly, Y. Racial/Ethnic, Socioeconomic, and Geographic Disparities in the Epidemiology of Knee and Hip Osteoarthritis. *Rheum. Dis. Clin. North. Am.* **2021**, *47*, 1–20. [CrossRef] [PubMed]
5. Hawker, G.A. Osteoarthritis is a serious disease. *Clin. Exp. Rheumatol.* **2019**, *120*, 3–6.
6. Briani, R.V.; Ferreira, A.S.; Pazzinatto, M.F.; Pappas, E.; De Oliveira Silva, D.; Azevedo, F.M. What interventions can improve quality of life or psychosocial factors of individuals with knee osteoarthritis? A systematic review with meta-analysis of primary outcomes from randomised controlled trials. *Br. J. Sports. Med.* **2018**, *52*, 1031–1038. [CrossRef]
7. Hamood, R.; Tirosh, M.; Fallach, N.; Chodick, G.; Eisenberg, E.; Lubovsky, O. Prevalence and Incidence of Osteoarthritis: A Population-Based Retrospective Cohort Study. *J. Clin. Med.* **2021**, *10*, 4282. [CrossRef]
8. Vitaloni, M.; Botto-van Bemden, A.; Sciortino Contreras, R.M.; Scotton, D.; Bibas, M.; Quintero, M.; Monfort, J.; Carné, X.; de Abajo, F.; Oswald, E.; et al. Global management of patients with knee osteoarthritis begins with quality of life assessment: A systematic review. *BMC Musculoskelet. Disord.* **2019**, *20*, 493. [CrossRef]
9. Fonseca-Rodrigues, D.; Rodrigues, A.; Martins, T.; Pinto, J.; Amorim, D.; Almeida, A.; Pinto-Ribeiro, F. Correlation between pain severity and levels of anxiety and depression in osteoarthritis patients: A systematic review and meta-analysis. *Rheumatology* **2021**, *61*, 53–75. [CrossRef]
10. Lee, Y.; Lee, S.H.; Lim, S.M.; Baek, S.H.; Ha, I.H. Mental health and quality of life of patients with osteoarthritis pain: The sixth Korea National Health and Nutrition Examination Survey (2013–2015). *PLoS ONE* **2020**, *15*, e0242077. [CrossRef]
11. Calders, P.; Van Ginckel, A. Presence of comorbidities and prognosis of clinical symptoms in knee and/or hip osteoarthritis: A systematic review and meta-analysis. *Semin. Arthritis. Rheum.* **2018**, *47*, 805–813. [CrossRef]
12. Nazarinasab, M.; Motamedfar, A.; Moqadam, A.E. Investigating mental health in patients with osteoarthritis and its relationship with some clinical and demographic factors. *Reumatologia* **2017**, *55*, 183–188. [CrossRef] [PubMed]
13. Kemp, A.H.; Tree, J.; Gracey, F.; Fisher, Z. Editorial: Improving Wellbeing in Patients with Chronic Conditions: Theory, Evidence, and Opportunities. *Front. Psychol.* **2022**, *13*, 868810. [CrossRef]
14. Kolasinski, S.L.; Neogi, T.; Hochberg, M.C.; Oatis, C.; Guyatt, G.; Block, J.; Callahan, L.; Copenhaver, C.; Dodge, C.; Felson, D.; et al. 2019 American College of Rheumatology/Arthritis Foundation Guideline for the Management of Osteoarthritis of the Hand, Hip, and Knee. *Arthritis. Care. Res.* **2020**, *72*, 149–162. [CrossRef] [PubMed]
15. Li, H.; George, D.M.; Jaarsma, R.L.; Mao, X. Metabolic Syndrome and Components Exacerbate Osteoarthritis Symptoms of Pain, Depression and Reduced Knee Function. *Ann. Transl. Med.* **2016**, *4*, 133. [CrossRef] [PubMed]
16. Moretti, L.; Coviello, M.; Rosso, F.; Calafiore, G.; Monaco, E.; Berruto, M.; Solarino, G. Current Trends in Knee Arthroplasty: Are Italian Surgeons Doing What Is Expected? *Medicina* **2022**, *58*, 1164. [CrossRef]
17. Toledano-González, A.; Labajos-Manzanares, T.; Romero-Ayuso, D. Well-Being, Self-Efficacy and Independence in older adults: A Randomized Trial of Occupational Therapy. *Arch. Gerontol. Geriatr.* **2019**, *83*, 277–284. [CrossRef]
18. Stoica, C.I.; Nedelea, G.; Cotor, D.C.; Gherghe, M.; Georgescu, D.E.; Dragosloveanu, C.; Dragosloveanu, S. The Outcome of Total Knee Arthroplasty for Patients with Psychiatric Disorders: A Single-Center Retrospective Study. *Medicina* **2022**, *58*, 1277. [CrossRef]
19. Pinto, S.; Fumincelli, L.; Mazzo, A.; Caldeira, S.; Martins, J.C. Comfort, well-being and quality of life: Discussion of the differences and similarities among the concepts. *Porto. Biomed. J.* **2017**, *2*, 6–12. [CrossRef]
20. RAND Health Care. 36-Item Short Form Survey (SF-36). Available online: https://www.rand.org/health-care/surveys_tools/mos/36-item-short-form.html (accessed on 2 August 2021).
21. Hejdysz, K.; Goślińska, J.; Wareńczak, A.; Dudzińska, J.; Adamczyk, E.; Sip, P.; Gośliński, J.; Owczarek, P.; Woźniak, A.; Lisiński, P. Manual Therapy Versus Closed Kinematic Exercises—The Influence on the Range of Movement in Patients with Knee Osteoarthritis: A Pilot Study. *Appl. Sci.* **2020**, *10*, 8605. [CrossRef]

22. Muntner, P.; Shimbo, D.; Carey, R.M.; Charleston, J.B.; Gaillard, T.; Misra, S.; Myers, M.G.; Ogedegbe, G.; Schwartz, J.E.; Townsend, R.R.; et al. Measurement of Blood Pressure in Humans: A Scientific Statement from the American Heart Association. *Hypertension* **2019**, *73*, e35–e66. [CrossRef]

23. International Diabetes Federation (IDF). IDF Consensus Worldwide Definition of the Metabolic Syndrome. Available online: https://www.idf.org/e-library/consensus-statements/60-idfconsensus-worldwide-definitionof-the-metabolic-syndrome. html (accessed on 22 August 2021).

24. van Dijk, G.M.; Veenhof, C.; Schellevis, F.; Hulsmans, H.; Bakker, J.P.; Arwert, H.; Dekker, J.H.; Lankhorst, G.J.; Dekker, J. Comorbidity, limitations in activities and pain in patients with osteoarthritis of the hip or knee. *BMC Musculoskelet. Disord.* **2008**, *9*, 95. [CrossRef] [PubMed]

25. Legha, A.; Burke, D.L.; Foster, N.E.; van der Windt, D.A.; Quicke, J.G.; Healey, E.L.; Runhaar, J.; Holden, M.A. Do comorbidities predict pain and function in knee osteoarthritis following an exercise intervention, and do they moderate the effect of exercise? Analyses of data from three randomized controlled trials. *Musculoskeletal. Care* **2020**, *18*, 3–11. [CrossRef] [PubMed]

26. Kim, S.K.; Choe, J.Y. Comorbidities and Health-Related Quality of Life in Subjects with Spine Osteoarthritis at 50 Years of Age or Older: Data from the Korea National Health and Nutrition Examination Survey. *Medicina* **2022**, *58*, 126. [CrossRef] [PubMed]

27. Mikic, A.; Minakovic, I.; Zvekic-Svorcan, J.; Jankovic, T.; Glomazic, H.; Boškovic, K. Predictors of emotional well-being in female patients with knee osteoarthritis. In Proceedings of the World Congress on Osteoporosis, Osteoarthritis and Musculoskeletal Diseases, Barcelona, Spain, 4–7 May 2023.

28. Lee, B.J.; Yang, S.; Kwon, S.; Choi, K.H.; Kim, W. Association between metabolic syndrome and knee osteoarthritis: A cross-sectional nationwide survey study. *J. Rehabil. Med.* **2019**, *51*, 464–470. [CrossRef]

29. Witjes, S.; van Geenen, R.C.; Koenraadt, K.L.; van der Hart, C.P.; Blankevoort, L.; Kerkhoffs, G.M.; Kuijer, P.P. Expectations of younger patients concerning activities after knee arthroplasty: Are we asking the right questions? *Qual. Life Res.* **2017**, *26*, 403–417. [CrossRef]

30. Giesbrecht, E.M.; Miller, W.C. A randomized control trial feasibility evaluation of an mHealth intervention for wheelchair skill training among middle-aged and older adults. *PeerJ* **2017**, *5*, e3879. [CrossRef]

31. Hyodo, K.; Masuda, T.; Aizawa, J.; Jinno, T.; Morita, S. Hip, knee, and ankle kinematics during activities of daily living: A cross-sectional study. *Braz. J. Phys. Ther.* **2017**, *21*, 159–166. [CrossRef]

32. Laires, P.A.; Laíns, J.; Miranda, L.C.; Cernadas, R.; Rajagopalan, S.; Taylor, S.D.; Silva, J.C. Inadequate pain relief among patients with primary knee osteoarthritis. *Rev. Bras. Reumatol. Engl. Ed.* **2017**, *57*, 229–237. [CrossRef]

33. Siviero, P.; Veronese, N.; Smith, T.; Stubbs, B.; Limongi, F.; Zambon, S.; Dennison, E.M.; Edwards, M.; Cooper, C.; Timmermans, E.J.; et al. Association Between Osteoarthritis and Social Isolation: Data from the EPOSA Study. *J. Am. Geriatr. Soc.* **2020**, *68*, 87–95. [CrossRef]

34. Bowden, J.L.; Callahan, L.F.; Eyles, J.P.; Kent, J.L.; Briggs, A.M. Realizing Health and Well-being Outcomes for People with Osteoarthritis Beyond Health Service Delivery. *Clin. Geriatr. Med.* **2022**, *38*, 433–448. [CrossRef] [PubMed]

35. Zheng, S.; Tu, L.; Cicuttini, F.; Zhu, Z.; Han, W.; Antony, B.; Wluka, A.E.; Winzenberg, T.; Aitken, D.; Blizzard, L.; et al. Depression in patients with knee osteoarthritis: Risk factors and associations with joint symptoms. *BMC Musculoskelet. Disord.* **2021**, *22*, 40. [CrossRef] [PubMed]

36. Timmermans, E.J.; de Koning, E.J.; van Schoor, N.M.; van der Pas, S.; Denkinger, M.D.; Dennison, E.M.; Maggi, S.; Pedersen, N.L.; Otero, Á.; Peter, R.; et al. Within-person pain variability and physical activity in older adults with osteoarthritis from six European countries. *BMC Musculoskelet. Disord.* **2019**, *20*, 12. [CrossRef] [PubMed]

37. Corcoran, M.P.; Nelson, M.E.; Sacheck, J.M.; Reid, K.F.; Kirn, D.; Fielding, R.A.; Folta, S.C. Recruitment of Mobility Limited Older Adults into a Facility-Led Exercise-Nutrition Study: The Effect of Social Involvement. *Gerontologist* **2016**, *56*, 669–676. [CrossRef]

38. Ali, S.A.; Kokorelias, K.M.; MacDermid, J.C.; Kloseck, M. Education and Social Support as Key Factors in Osteoarthritis Management Programs: A Scoping Review. *Arthritis* **2018**, 2496190. [CrossRef] [PubMed]

![MDPI]

Article

Immersion Ultrasound Therapy in Combination with Manual Therapy in the Treatment of Ischemic Digital Ulcers in Systemic Sclerosis

Dalila Scaturro [1,*,†], **Antimo Moretti** [2], **Fabio Vitagliani** [3,†], **Giuliana Guggino** [4], **Sofia Tomasello** [5], **Davide Lo Nardo** [3], **Lorenza Lauricella** [6], **Giovanni Iolascon** [2] and **Giulia Letizia Mauro** [1]

[1] Department of Surgical, Oncological and Stomatological Disciplines, University of Palermo, 90127 Palermo, Italy; giulia.letiziamauro@unipa.it

[2] Department of Medical and Surgical Specialties and Dentistry, University of Campania "Luigi Vanvitelli", 80138 Naples, Italy; antimo.moretti@unicampania.it (A.M.); giovanni.iolascon@unicampania.it (G.I.)

[3] Faculty of Medicine and Surgery, University of Catania, 90121 Catania, Italy; fabiovitagliani93@libero.it (F.V.); davidelon.991@gmail.com (D.L.N.)

[4] Rheumatology Section, Biomedical Department of Internal Medicine, University Hospital "P. Giaccone", 90127 Palermo, Italy; giuliana.guggino@unipa.it

[5] Faculty of Medicine and Surgery, University of Palermo, 90100 Palermo, Italy; sofia.tomasello@hotmail.com

[6] Paolo Giaccone University Hospital, 90127 Palermo, Italy; lorenzalauricella@libero.it

* Correspondence: dalila.scaturro@unipa.it; Tel.: +39-3206-945-411

† These authors contributed equally to this work.

Abstract: *Background and Objectives*: Digital ulcers (DUs) are the most common complication in patients with Systemic Sclerosis (SSc). They cause pain with hand dysfunction and negatively impact activities of daily and working life. Our study aims to evaluate the efficacy of a combined treatment of manual therapy and ultrasound therapy in SSc patients with ischemic DU (IDU) compared to manual therapy alone. *Materials and Methods*: We conducted a before-and-after study (non-randomized study). We enrolled a consecutive series of IDU patients undergoing rehabilitation treatment and divided them into two groups: a treatment group consisting of patients undergoing a combination of manual therapy and US water immersion and a standard care group consisting of patients subjected to manual therapy alone. At the time of the first visit (T0) and at the end of the 4-week rehabilitation period (T1), we evaluated functional capacity, pain intensity, ulcer evolution, and quality of life. *Results*: In the treatment group, we observed a statistically significant improvement in the functional capacity of the hand (DHI: 28.15 ± 11.0 vs. 19.05 ± 8.83; $p < 0.05$), pain (NRS: 5.55 ± 1.2 vs. 2.9 ± 1.09; $p < 0.05$), and PSST score (24.4 ± 4.0 vs. 16.2 ± 2.36; $p < 0.05$). In the standard care group, we observed a statistically significant improvement only for the functional capacity of the hand (DHI: 28.85 ± 9.72 vs. 22.7 ± 7.68; $p < 0.05$). Finally, from the comparison between the treatment group and the standard care group, we observed statistically significant improvements in pain (2.9 ± 1.09 vs. 4.5 ± 1.07; $p < 0.05$) and in the PSST scale (16.2 ± 2.36 vs. 20.4 ± 4.02; $p < 0.05$). Furthermore, at the end of treatment in the treatment group, 15 ulcers (62.5%) were completely healed, while in the standard care group, only 3 ulcers were completely healed (14.3%). *Conclusions*: Combined treatment with manual therapy and ultrasound therapy appears to be useful in the management of IDU in patients with scleroderma.

Keywords: systemic sclerosis; rehabilitation; skin ulcer; ultrasound therapy; disability

Citation: Scaturro, D.; Moretti, A.; Vitagliani, F.; Guggino, G.; Tomasello, S.; Lo Nardo, D.; Lauricella, L.; Iolascon, G.; Letizia Mauro, G. Immersion Ultrasound Therapy in Combination with Manual Therapy in the Treatment of Ischemic Digital Ulcers in Systemic Sclerosis. *Medicina* **2023**, *59*, 1335. https://doi.org/10.3390/medicina59071335

Academic Editor: Dejan Nikolic

Received: 23 June 2023
Revised: 17 July 2023
Accepted: 18 July 2023
Published: 20 July 2023

1. Introduction

Systemic sclerosis (SSc) is a chronic autoimmune disease characterized by vascular disease, inflammation, immune response, and accumulation of collagen in the skin and other organs resulting in fibrosis [1,2]. This condition leads to an alteration of tissue architecture with consequent loss of organ function in the terminal stages of the disease [3].

The etiology is still uncertain, but an important role in its development seems to be played by genetic and environmental factors [2]. It has an incidence of between 4 and 43 million people per year and mostly affects women [4–7].

According to the new 2017 criteria of the SSc classification of the American College of Rheumatology (ACR) and the European League Against Rheumatism (EULAR) [8], skin thickening of the fingers is sufficient to make a diagnosis. In addition, there are seven other typical pathological conditions, including fingertip ulcers or pitting scars, localized finger telangiectasias, capilloscopic changes, pulmonary hypertension associated with digital ulcers (DUs), Raynaud's phenomenon, and the presence of autoantibodies [8,9].

Thickening and fibrosis of the skin and internal organs, caused by progressive vascular disease, are the main clinical manifestations [2,3]. Involvement of the hands is responsible for a marked disability in these patients due to the presence of swollen fingers, DUs, cutaneous sclerosis, calcinosis of the skin, and pruritus [10,11].

DUs are the most common complication, affecting over 50–70% of patients with SSc. They are recurring lesions affecting hands and feet, characterized by loss of continuity and depth of the skin. They are caused by ischemic vessel damage and are indicators of overall disease severity and organ involvement [10,12]. DUs often lead to pain with hand dysfunction and have a negative impact on activities of daily living and work, resulting in disability [10]. Furthermore, the presence of DUs requires continuous follow-up to prevent possible complications, such as infections that can involve both the skin and underlying tissues, such as bone, resulting in osteomyelitis [12,13].

Depending on the extent of skin involvement, systemic sclerosis can be divided into a limited cutaneous form (lcSSc), with involvement limited to the skin of the hands, forearms, feet, and face, and a diffuse cutaneous form (dcSSc), involving skin proximal to the elbows and trunk [1].

Currently, the treatment of SSc focuses on various pharmacological and non-pharmacological interventions based on recommendations and evidence published by the EULAR and EULAR Scleroderma Trials and Research (EUSTAR) groups [8]. To date, there are no specific treatments for the treatment of DUs. Several systemic treatments are available [14–19], such as immunomodulators [8], UVA phototherapy [16], topical calcitriol [17], and retinoids [18]. In addition, local therapies with good aesthetic and functional results can also be used, such as the injection of autologous fat and platelet-rich plasma (PRP) [19,20] or adipose-derived stromal cells (ADSC) combined with hyaluronic acid (HA) [21] or HA and PRP [21,22]. Finally, physical modalities for DUs have also been proposed, including ultrasound therapy [23–25] and connective tissue massage, mainly using the McMennel technique [26].

Our study aims to evaluate the efficacy of ultrasound therapy in combination with manual therapy in the management of systemic sclerosis patients with IDU in terms of hand functional capacity, pain, wound healing, and quality of life.

2. Materials and Methods

This is a before-and-after study (non-randomized trial) conducted at the Functional Recovery and Rehabilitation Unit of the A.O.U.P. Paolo Giaccone in Palermo, in collaboration with the Rheumatology Unit of the same hospital. Patients with scleroderma presenting IDU were consecutively enrolled (from April 2022 to November 2022).

All aspects of the study were reviewed and approved by the local "Palermo 1" ethics committee, with reference number 6/2020. The ethical guidelines of the Declaration of Helsinki were followed for the study, and the information was handled following the guidelines of the Good Clinical Practice (GCP). The study was registered with ClinicalTrials.gov (NCT05907200).

2.1. Patients

Inclusion criteria were a diagnosis of SSc according to the ACR and EULAR criteria [8], the presence of IDU in the active phase, naïve to rehabilitation treatment for their hands and upper limbs, and capacity to provide informed consent. In both groups, all patients continued their usual pharmacological treatments (alprostadil-α-cyclodextran, calcium channel blockers, topical glyceryl trinitrate, proton pump inhibitors, cleobopride, steroids, cyclophosphamide, azathioprine, D-penicillamine, and methotrexate).

Exclusion criteria were the presence of skin lesions due to other conditions (e.g., trauma), pregnancy, infectious diseases (e.g., HIV, HBV, HCV), myositis, arthritis, other rheumatological diseases, and immunodepression.

Using our hospital's database, we enrolled a consecutive series of patients with IDU who had undergone rehabilitation treatment and met our inclusion criteria.

2.2. Intervention and Control

Based on the type of rehabilitation treatment received, the patients were divided into two groups: a treatment group consisting of patients who underwent rehabilitation treatment consisting of a combination of manual therapy and water immersion US and a standard care group consisting of patients who underwent rehabilitation treatment consisting of manual therapy alone [26]. The rehabilitation treatment was performed by all patients at the same time of year (from April to June) and was carried out for 20 sessions daily, 5 days a week, for 4 consecutive weeks. Some clinical information and educational recommendations on skin nutrition and protection were provided to all patients.

2.3. Description of Manual Therapy Techniques

The proposed manual therapy lasted 90 min and involved a combination of three different techniques: McMennel manipulation, connective tissue massage, and the Pompage mobilization technique [26].

McMennel manipulation is a technique for reducing joint stiffness and reducing pain by stretching the capsuloligo-mental complex. It was used for 40 min. It prevents the development of claw deformity and increases the trophism of the cartilage, resulting in an improvement in hand mobility and the extrinsic strength of the hand muscles, decreasing joint pain and stiffness [27].

Connective tissue massage (lasting 30 min) improves the blood circulation of the skin and plays a muscle-relaxing role, keeping the skin soft [28].

The Pompage mobilization technique lasts 20 min and consists of slow and progressive mobilizations through a rhythmic and regular movement of traction and release, which allows the recovery of the physiological length of the soft tissues [28].

2.4. Description of Therapeutic US Technique

Patients in the treatment group also received treatment with a US (UT2 CE0476 certified I-Tech medical device). The USs were used with a frequency of 1 MHz, an intensity of 1 W/cm^2, a duty cycle of 60%, and a duration of 15 min per session. The patient's hands were immersed inside a metal container with a diameter of 90 cm, containing 4 L of water at a temperature of 37–37.5 °C. An emitter handpiece with radiant air of 5 cm^2 was then inserted inside the container, about 2 cm from the body surface. The action of therapeutic ultrasound reduces inflammation and bacterial counts on wounds and improves cell proliferation and neoangiogenesis [23,24].

2.5. Outcome Measures

Demographic, anthropometric, and clinical data were collected at baseline. Height and weight were measured in each subject, and BMI was then calculated. At the time of the first visit (T0) and the end of the 4 weeks of rehabilitation (T1), the same expert physiatrist (D.S.) subjected the patients to the following evaluation scales: numerical evaluation scale (NRS) [29] for ache; Duruoz hand index (DHI) [30] for functional ability; Pressure Sore Status Tool (PSST) [31] for ulcer healing; and Short Form Health Survey 36 (SF-36) scales [32] for quality of life (QoL). Our primary outcome measure was functional capacity (DHI), while our secondary outcome measures were pain (NRS), ulcer healing (PSST), and QoL (SF-36).

The DHI is a self-reported questionnaire designed to assess hand activity limits. It includes 18 items that are rated on a 6-point scale where 0 is "without difficulty", and 5 is "impossible" [30]. The PSST includes 13 items relating to the characteristics of the wound and surrounding tissue, which are assigned a score from 1 to 5. The total sum of all scores provides the PSST score, where the higher the final scores, the more severe the state of the lesion will be [31]. The SF-36 is a questionnaire comprising eight multiple-choice questions that can be divided into two large subgroups: the physical component of the disease and the mental component of the disease. A score is assigned to each scale; the higher the score, the better the state of mental and physical health. The score ranges from 0 (worst state of health) to 100 (best state of health) [32].

2.6. Statistical Analysis

All analyzes were performed using R software (R Core Team, 2013). The sample size was calculated to detect a difference in DHI of 5.1 units and a loss of follow-up of 20%. The estimated sample size calculation was 24 for each group. For the comparison of the quantitative variables, we used the t-test, and for the ordinal variables, Mood's median test. Values < 0.05 were considered statistically significant.

3. Results

Sixty-two scleroderma patients with IDU were considered in this study. Of these, 7 had not completed the proposed rehabilitation treatment, and 10 did not present themselves for follow-up at the end of the 4-week rehabilitation treatment. A total of 45 patients were included in the study, 24 patients belonging to the treatment group and 21 patients belonging to the standard care group (Figure 1).

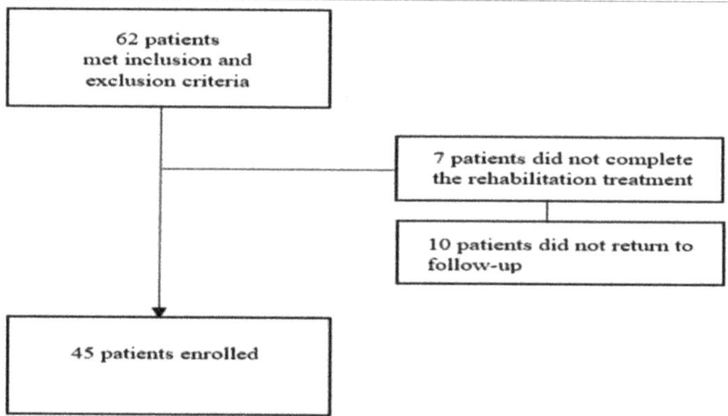

Figure 1. Patient recruitment.

The included patients had a mean age of 61.12 ± 8.83 years and included 33 (73.3%) women and 12 (26.7%) men, with a mean BMI of 24.6 ± 2.11 kg/m². Thirteen patients (32.5%) had pulmonary involvement (interstitial lung disease and/or pulmonary hypertension). All patients had ulcers on their fingertips and presented stiffness and loss of joint function due to flexion contractures caused by skin retraction. The included patients had a mean DHI value of 28.5 ± 10.4, with a mean perceived pain of 5.52 ± 1.22 points on the NRS. Finally, the mean value of the PSST scale was 24.32 ± 4.14, and of the SF-36 scale, 57.2 ± 7.98. No statistically significant between-group difference in baseline characteristics was reported (Table 1).

Table 1. General characteristics at baseline.

Characteristics	Total (*n* = 45)	Treatment Group (*n* = 24)	Standard Care Group (*n* = 21)	*p*-Value
Age, mean ± SD	61.12 ± 8.83	61.05 ± 9.3	61.2 ± 8.34	0.96
Sex, n° (%)				
Male	12 (26.7)	7 (29.1)	5 (23.8)	0.88
Female	33 (73.3)	17 (70.9)	16 (76.2)	
BMI (kg/m^2), mean ± SD	24.6 ± 2.11	24.3 ± 1.88	25.1 ± 1.55	0.56
DHI, mean±SD	28.5 ± 10.4	28.15 ± 11.0	28.85 ± 9.72	0.82
NRS, mean ± SD	5.52 ± 1.22	5.55 ± 1.2	5.5 ± 1.24	0.89
PSST, mean ± SD	24.32 ± 4.14	24.4 ± 4.0	24.25 ± 4.27	0.90
SF-36, mean ± SD	57.2 ± 7.98	57.05 ± 9.1	57.35 ± 6.66	0.65

DHI: Duruoz Hand Index; NRS: Numerical evaluation scale; PSST: Pressure Sore Status Tool; SF-36: Short Form Health Survey 36.

Table 2 shows the effects of the combined rehabilitation treatment of immersion ultrasound therapy and manual therapy in the treatment group at T1. We observed statistically significant improvements in functional capacity (28.15 ± 11.0 vs. 19.05 ± 8.83; $p < 0.05$), pain (5.55 ± 1.2 vs. 2.9 ± 1.09; $p < 0.05$), and PSST score (24.4 ± 4.0 vs. 16.2 ± 2.36; $p < 0.05$). Finally, the QoL did not significantly change at follow-up (SF-36: 57.05 ± 9.1 vs. 52.0 ± 8.75; $p = 0.08$) (Table 2).

Table 2. Effect of combined treatment of immersion ultrasound therapy and manual therapy in IDUs in the treatment group.

Characteristics	T0	T1	*p*-Value
DHI, mean ± SD	28.15 ± 11.0	19.05 ± 8.83	<0.05
NRS, mean ± SD	5.55 ± 1.2	2.9 ± 1.09	<0.05
PSST, mean ± SD	24.4 ± 4.0	16.2 ± 2.36	<0.05
SF-36, mean ± SD	57.05 ± 9.1	52.0 ± 8.75	0.08

DHI: Duruoz Hand Index; NRS: Numerical evaluation scale; PSST: Pressure Sore Status Tool; SF-36: Short Form Health Survey 36.

Table 3 shows the effects of manual therapy alone in the standard care group at T1. A statistically significant improvement was observed only for the functional capacity of the hands (28.85 ± 9.72 vs. 23.7 ± 7.68; $p < 0.05$). No statistically significant improvement was observed for pain (5.5 ± 1.24 vs. 4.5 ± 1.07; $p = 0.08$), PSST scale (24.25 ± 4.27 vs. 20.4 ± 4.02; $p = 0.16$), and quality of life (57.35 ± 6.66 vs. 54.5 ± 6.54; $p = 0.18$) (Table 3).

Table 3. Effect of manual therapy in IDUs in the standard care group.

Characteristics	T0	T1	*p*-Value
DHI, mean ± SD	28.85 ± 9.72	23.7 ± 7.68	<0.05
NRS, mean ± SD	5.5 ± 1.24	4.5 ± 1.07	0.08
PSST, mean ± SD	24.25 ± 4.27	20.4 ± 4.02	0.16
SF-36, mean ± SD	57.35 ± 6.66	54.5 ± 6.54	0.18

DHI: Duruoz Hand Index; NRS: Numerical evaluation scale; PSST: Pressure Sore Status Tool; SF-36: Short Form Health Survey 36.

Finally, in Table 4, we compare the results obtained in the treatment group and the standard care group at T1. From the comparison, in the treatment group compared to the standard care group, we observed statistically significant improvements in pain (2.9 ± 1.09 vs. 4.5 ± 1.07; $p < 0.05$) and in the PSST scale (16.2 ± 2.36 vs. 20.4 ± 4.02; $p < 0.05$) (Table 4). Furthermore, at the end of the treatment in the treatment group, 15 ulcers (62.5%) were completely healed, while in the standard care group, only 3 ulcers were completely healed (14.3%).

Table 4. Comparison between the treatment group and the standard care group at T1.

Characteristics	Treatment Group	Standard Care Group	*p*-Value
DHI, mean ± SD	19.05 ± 8.83	23.7 ± 7.68	0.07
NRS, mean ± SD	2.9 ± 1.09	4.5 ± 1.07	<0.05
PSST, mean ± SD	16.2 ± 2.36	20.4 ± 4.02	<0.05
SF-36, mean ± SD	52.0 ± 8.75	54.5 ± 6.54	0.29

DHI: Duruoz Hand Index; NRS: Numerical evaluation scale; PSST: Pressure Sore Status Tool; SF-36: Short Form Health Survey 36.

4. Discussion

IDUs are common complications in patients with scleroderma [33]. They are very painful lesions with slow healing that lead to a great deal of disability in affected patients. It is estimated that 15–25% of patients with SSc have active IDUs [34]. In this study, we aimed to compare the effectiveness of two different rehabilitation treatments in the management of IDUs of patients with scleroderma in terms of functional capacity. In addition, the effects of these interventions on pain, ulcer healing, and QoL were also compared.

Our results showed a significant improvement in the functional capacity of the hand in both groups. This finding emphasizes the importance of using manual therapy in combination with pharmacological treatments for the management of IDU in patients with scleroderma. SSc patients with IDU have reduced hand mobility and increased global and hand disability compared to those without IDU [35]. Rehabilitation treatment appears useful in reducing the impact of hand impairments in patients with SSc. Boggi et al. [26] demonstrated that the combination of connective tissue massage and McMennell's joint manipulation proved to be more effective than a program based on daily exercises performed at home in treating the hands of patients with scleroderma. At the end of the treatment period, the authors observed an increase in the HAMIS test and the Cochin Hand Functional Disability Scale, as well as an improvement in mobility, fine movements, and hand function [26]. On the other hand, Mugii et al. [36] studied the effect of self-administered finger lengthening in patients with SSc, showing an improvement in the range of motion in each finger after as early as one month. The authors demonstrated that the combination of self-stretching exercises can be useful for maintaining the efficacy achieved with a rehabilitation program supervised by a physiotherapist. In a quasi-experimental study, the exercise demonstrated positive effects on isometric muscle strength, muscle function, and hand functional capacity in SSc-related IDU. However, it has shown little effect on the healing of IDU or Raynaud's phenomenon [37].

Another finding highlighted in our study was that significant pain relief was obtained only in the group that received a combination of manual therapy and therapeutic US, compared to patients treated with manual therapy alone. Pain management is of paramount importance in patients with IDUs. Moreover, in patients with systemic sclerosis, non-steroidal anti-inflammatory drugs (NSAIDs), despite their efficacy, should be avoided in favor of paracetamol and opiates due to their known vascular effects. The cause of pain in IDUs is tissue ischemia [35], and the pain relief observed in our study could result from the thermal effect of US. This method, by means of mechanical vibrations, transfers energy to the tissues, inducing the dilation of blood vessels and increasing cell metabolism through the supply of oxygen and nutrients. In addition, the thermal effect of US also has an analgesic action, resulting in changes in pain threshold and tissue viscoelasticity [38,39].

Our study also suggests that ulcer healing seems to be better in the group of patients undergoing the combined treatment of manual therapy and therapeutic US, compared with the standard care group. In addition to the vasodilation induced by US, the mechanical action of this intervention might positively affect the wound tissue by promoting and stimulating cell proliferation [40]. This finding is in line with a previous case report of a patient with a large, painful, and infected DU that was treated with low-intensity US three times a week for 5 min. After 8 weeks from the start of treatment, a decrease in pain score from 10 to 0 was observed, and complete wound closure after 10 weeks [40].

Finally, although a greater improvement in SF-36 score was observed in the treatment group, no significant improvement in quality of life was reported in any group. However, this finding should not be surprising as the main difficulties patients with systemic sclerosis complain about are in activities of daily living (ADLs), including walking, cleaning, and sports training [36]. Hence, although most ADLs involve the use of the hands, the major limitations in these activities are likely related to other disease-related complications [36]. The strength of our study is that it is the first comparison between two different rehabilitation methods in the management of IDU. Our previous study focused on evaluating the effectiveness of the combined rehabilitation treatment of manual therapy and therapeutic US on patients with IDU but lacked a standard care group. However, there are several limitations to this study. First, the study design increases the risk of bias. Furthermore, two further limitations are the short follow-up period which, however, would have been influenced by seasonality, the lack of assessment of the frequency of relapses, and the efficacy of the treatment over time.

Furthermore, we did not make the comparison with a control group because the ethics of our department requires that all subjects with IDU undergo therapy. Finally, a rare disease causes a small sample.

5. Conclusions

In patients with systemic sclerosis, the association between ultrasound and manual therapy can be considered effective treatment for IDU, pain, and hand motility. It can be said that ultrasound therapy is safe even if there are ulcers. Manual therapy is essential to improve motility and for carrying out activities.

Author Contributions: Conceptualization, D.S. and F.V.; methodology, F.V.; software, S.T.; validation, G.I., G.L.M. and D.S.; formal analysis, F.V.; investigation, D.S. and F.V.; data curation, D.L.N. and L.L.; writing—original draft preparation, F.V. and S.T.; review A.M. and G.G.; supervision, A.M.; G.G. and G.L.M. All authors have read and agreed to the published version of the manuscript.

Funding: This research received no external funding.

Institutional Review Board Statement: The study was conducted in accordance with the Declaration of Helsinki and approved Local Ethics Committee "Palermo 1" (protocol code 06/2020 on 15 June 2020).

Informed Consent Statement: Written informed consent has been obtained from the patient(s) to publish this paper.

Data Availability Statement: Data used to support the findings of this study are available from the corresponding author upon request.

Conflicts of Interest: The authors declare no conflict of interest.

References

1. Hinchcliff, M.; Varga, J. Systemic sclerosis/scleroderma: A treatable multisystem disease. *Am. Fam. Physician* **2008**, *78*, 961–968. [PubMed]
2. Di Benedetto, P.; Ruscitti, P.; Liakouli, V.; Cipriani, P.; Giacomelli, R. The Vessels Contribute to Fibrosis in Systemic Sclerosis. *Isr. Med. Assoc. J.* **2019**, *21*, 471–474. [PubMed]
3. Ho, Y.Y.; Lagares, D.; Tager, A.M.; Kapoor, M. Fibrosis—A lethal component of systemic sclerosis. *Nat. Rev. Rheumatol.* **2014**, *10*, 390–402. [CrossRef] [PubMed]
4. Hoffmann-Vold, A.M.; Midtvedt, Ø.; Molberg, Ø.; Garen, T.; Gran, J.T. Prevalence of systemic sclerosis in south-east Norway. *Rheumatology* **2012**, *51*, 1600–1605. [CrossRef] [PubMed]
5. Kaipiainen-Seppänen, O.; Aho, K. Incidence of rare systemic rheumatic and connective tissue diseases in Finland. *J. Intern. Med.* **1996**, *240*, 81–84. [CrossRef] [PubMed]
6. Bernatsky, S.; Joseph, L.; Pineau, C.A.; Belisle, P.; Hudson, M.; Clarke, A.E. Scleroderma prevalence: Demographic variations in a population-based sample. *Arthritis Rheum.* **2009**, *61*, 400–404. [CrossRef]
7. Nikpour, M.; Stevens, W.M.; Herrick, A.L.; Proudman, S.M. Epidemiology of systemic sclerosis. *Best. Pract. Res. Clin. Rheumatol.* **2010**, *24*, 857–869. [CrossRef]

8. Kowal-Bielecka, O.; Fransen, J.; Avouac, J.; Becker, M.; Kulak, A.; Allanore, Y.; Distler, O.; Clements, P.; Cutolo, M.; Czirjak, L.; et al. Update of EULAR recommendations for the treatment of systemic sclerosis. *Ann. Rheum. Dis.* **2017**, *76*, 1327–1339. [CrossRef]
9. Denton, C.P.; Khanna, D. Systemic sclerosis. *Lancet* **2017**, *390*, 1685–1699. [CrossRef]
10. Hughes, M.; Bruni, C.; Ruaro, B.; Confalonieri, M.; Matucci-Cerinic, M.; Bellando-Randone, S. Digital Ulcers in Systemic Sclerosis. *Presse Med.* **2021**, *50*, 104064. [CrossRef]
11. Abraham, S.; Steen, V. Optimal management of digital ulcers in systemic sclerosis. *Ther. Clin. Risk Manag.* **2015**, *15*, 939–947.
12. Starnoni, M.; Pappalardo, M.; Spinella, A.; Testoni, S.; Lattanzi, M.; Feminò, R.; De Santis, G.; Salvarani, C.; Giuggioli, D. Systemic sclerosis cutaneous expression: Management of skin fibrosis and digital ulcers. *Ann. Med. Surg.* **2021**, *71*, 102984. [CrossRef]
13. Poole, J.L.; Macintyre, N.J.; Deboer, H.N. Evidence-based management of hand and mouth disability in a woman living with diffuse systemic sclerosis (scleroderma). *Physiother Can.* **2013**, *65*, 317–320. [CrossRef]
14. Badea, I.; Taylor, M.; Rosenberg, A.; Foldvari, M. Pathogenesis and therapeutic approaches for improved topical treatment in localized scleroderma and systemic sclerosis. *Rheumatology* **2009**, *48*, 213–221. [CrossRef]
15. Kreuter, A.; Hyun, J.; Stücker, M.; Sommer, A.; Altmeyer, P.; Gambichler, T. A randomized controlled study of low-dose UVA1, medium-dose UVA1, and narrowband UVB phototherapy in the treatment of localized scleroderma. *J. Am. Acad. Dermatol.* **2006**, *54*, 440–447. [CrossRef]
16. Elst, E.F.; Van Suijlekom-Smit, L.W.; Oranje, A.P. Treatment of linear scleroderma with oral 1,25-dihydroxyvitamin D3 (calcitriol) in seven children. *Pediatr. Dermatol.* **1999**, *16*, 53–58. [CrossRef]
17. Mizutani, H.; Yoshida, T.; Nouchi, N.; Hamanaka, H.; Shimizu, M. Topical tocoretinate improved hypertrophic scar, skin sclerosis in systemic sclerosis and morphea. *J. Dermatol.* **1999**, *26*, 11–17. [CrossRef]
18. Gheisari, M.; Ahmadzadeh, A.; Nobari, N.; Iranmanesh, B.; Mozafari, N. Autologous Fat Grafting in the Treatment of Facial Scleroderma. *Dermatol. Res. Pract.* **2018**, *2018*, 6568016. [CrossRef]
19. Griffin, M.F.; Almadori, A.; Butler, P.E. Use of Lipotransfer in Scleroderma. *Aesthet. Surg. J.* **2017**, *37*, S33–S37. [CrossRef]
20. Scuderi, N.; Ceccarelli, S.; Onesti, M.G.; Fioramonti, P.; Guidi, C.; Romano, F.; Frati, L.; Angeloni, A.; Marchese, C. Human adipose-derived stromal cells for cell-based therapies in the treatment of systemic sclerosis. *Cell Transplant.* **2013**, *22*, 779–795. [CrossRef]
21. Pirrello, R.; Verro, B.; Grasso, G.; Ruscitti, P.; Cordova, A.; Giacomelli, R.; Ciccia, F.; Guggino, G. Hyaluronic acid and platelet-rich plasma, a new therapeutic alternative for scleroderma patients: A prospective open-label study. *Arthritis Res. Ther.* **2019**, *21*, 286. [CrossRef] [PubMed]
22. Pirrello, R.; Schinocca, C.; Scaturro, D.; Rizzo, C.; Terrana, P.; Tumminelli, L.; Ruscitti, P.; Cordova, A.; Giacomelli, R.; Ciccia, F.; et al. Additional Treatment for Digital Ulcers in Patients with Systemic Sclerosis A Prospective Open-Label Multi-Arm Study for the use of Platelet Rich Plasma Lipofilling and Ultrasound-Based Treatments. *J. Biotechnol. Biomed.* **2021**, *4*, 84–107. [CrossRef]
23. Samuels, J.A.; Weingarten, M.S.; Margolis, D.J.; Zubkov, L.; Sunny, Y.; Bawiec, C.R.; Conover, D.; Lewin, P.A. Low-frequency (<100 kHz), low-intensity (<100 mW/cm2) ultrasound to treat venous ulcers: A human study and in vitro experiments. *J. Acoust. Soc. Am.* **2013**, *134*, 1541–1547. [PubMed]
24. Lai, J.; Pittelkow, M.R. Physiological effects of ultrasound mist on fibroblasts. *Int. J. Dermatol.* **2007**, *46*, 587–593. [CrossRef]
25. Scaturro, D.; Guggino, G.; Terrana, P.; Vitagliani, F.; Falco, V.; Cuntrera, D.; Benedetti, M.G.; Moretti, A.; Iolascon, G.; Mauro, G.L. Rehabilitative interventions for ischaemic digital ulcers, pain, and hand functioning in systemic sclerosis: A prospective before-after study. *BMC Musculoskelet. Disord.* **2022**, *23*, 193. [CrossRef] [PubMed]
26. Maddali Bongi, S.; Del Rosso, A.; Galluccio, F.; Sigismondi, F.; Miniati, I.; Conforti, M.L.; Nacci, F.; Matucci Cerinic, M. Efficacy of connective tissue massage and McMennell joint manipulation in the rehabilitative treatment of the hands in systemic sclerosis. *Clin. Rheumatol.* **2009**, *28*, 1167–1173. [CrossRef]
27. Greenman, P.E. Mobilization with and without impulse technique. In *Principles of Manual Medicine*, 3rd ed.; Lippincott Williams & Wilkins: Philadelphia, PA, USA, 2003; pp. 107–112.
28. Alpiner, N.; Oh, T.H.; Hinderer, S.R.; Brander, V.A. Rehabilitation in joint and connective tissue diseases. 1. Systemic diseases. *Arch. Phys. Med. Rehabil.* **1995**, *75*, S32–S40. [CrossRef]
29. Hylands-White, N.; Duarte, R.V.; Raphael, J.H. An overview of treatment approaches for chronic pain management. *Rheumatol. Int.* **2017**, *37*, 29–42. [CrossRef]
30. Brower, L.M.; Poole, J.L. Reliability and validity of the Duruoz Hand Index in persons with systemic sclerosis (scleroderma). *Arthritis Rheum.* **2004**, *51*, 805–809, Correction in *Arthritis Rheum.* **2005**, *53*, 303. [CrossRef]
31. Sugihara, F.; Inoue, N.; Venkateswarathirukumara, S. Ingestion of bioactive collagen hydrolysates enhanced pressure ulcer healing in a randomized double-blind placebo-controlled clinical study. *Sci. Rep.* **2018**, *8*, 11403. [CrossRef]
32. Li, L.; Cui, Y.; Chen, S.; Zhao, Q.; Fu, T.; Ji, J.; Li, L.; Gu, Z. The impact of systemic sclerosis on health-related quality of life assessed by SF-36: A systematic review and meta-analysis. *Int. J. Rheum. Dis.* **2018**, *21*, 1884–1893. [CrossRef]
33. Korn, J.H.; Mayes, M.; Cerinic, M.M.; Rainisio, M.; Pope, J.; Hachulla, E.; Rich, E.; Carpentier, P.; Molitor, J.; Seibold, J.R.; et al. Digital ulcers in systemic sclerosis: Prevention by treatment with bosentan, an oral endothelin receptor antagonist. *Arthritis Rheum.* **2004**, *50*, 3985–3993. [CrossRef]
34. Schiopu, E.; Impens, A.J.; Phillips, K. Digital ischemia in scleroderma spectrum of diseases. *Int. J. Rheumatol.* **2010**, *2010*, 923743. [CrossRef]

35. Mouthon, L.; Carpentier, P.H.; Lok, C.; Clerson, P.; Gressin, V.; Hachulla, E.; Bérezné, A.; Diot, E.; Van Kien, A.K.; Jego, P.; et al. Ischemic digital ulcers affect hand disability and pain in systemic sclerosis. *J. Rheumatol.* **2014**, *41*, 1317–1323. [CrossRef]
36. Mugii, N.; Hamaguchi, Y.; Maddali-Bongi, S. Clinical significance and usefulness of rehabilitation for systemic sclerosis. *J. Scleroderma Relat. Disord.* **2018**, *3*, 71–80. [CrossRef]
37. Moran, M.E. Scleroderma and evidence based non-pharmaceutical treatment modalities for digital ulcers: A systematic review. *J. Wound Care* **2014**, *23*, 510–516. [CrossRef]
38. Abousenna, M.M.H.S.; Shalaby, A.S.M.; Chahal, A.; Shaphe, A. A comparison of low dose ultrasound and far-infrared therapies in patients with mechanical neck pain. *J. Pak. Med. Assoc.* **2021**, *71*, 397–401. [CrossRef]
39. Morishita, K.; Karasuno, H.; Yokoi, Y.; Morozumi, K.; Ogihara, H.; Ito, T.; Hanaoka, M.; Fujiwara, T.; Fujimoto, T.; Abe, K. Effects of therapeutic ultrasound on range of motion and stretch pain. *J. Phys. Ther. Sci.* **2014**, *26*, 711–715. [CrossRef]
40. Liguori, P.A.; Peters, K.L.; Bowers, J.M. Combination of negative pressure wound therapy and acoustic pressure wound therapy for treatment of infected surgical wounds: A case series. *Ostomy Wound Manag.* **2008**, *54*, 50–53.

Article

The Impact of Complex Rehabilitation Treatment on Sarcopenia—Pathology with an Endocrine Morphological Substrate and Musculoskeletal Implications

Liliana-Elena Stanciu [1,2], Mădălina-Gabriela Iliescu [1,2,*], Carmen Oprea [1,2], Elena-Valentina Ionescu [1,2], Adina Petcu [3], Viorela Mihaela Ciortea [4,*], Lucian Cristian Petcu [5], Sterian Apostol [1,2], Andreea-Dalila Nedelcu [1], Irina Motoașcă [4] and Laszlo Irsay [4]

[1] Balneal and Rehabilitation Sanatorium of Techirghiol, 34-40 Dr. Victor Climescu Street, 906100 Techirghiol, Romania; lilianastanciu77@yahoo.com (L.-E.S.); carmen_oprea_cta@yahoo.com (C.O.); elena_valentina_ionescu@yahoo.com (E.-V.I.); sterianapostol@yahoo.com (S.A.); dalilanedelcu@yahoo.ro (A.-D.N.)

[2] Department of Physical Medicine and Rehabilitation, Faculty of Medicine, "Ovidius" University of Constanta, 1 University Alley, Campus—Corp B, 900470 Constanta, Romania

[3] Faculty of Pharmacy, "Ovidius" University of Constanta, 1 University Alley, Campus—Corp B, 900470 Constanta, Romania; adinpetcu@yahoo.com

[4] Department of Rehabilitation Medicine, University of Medicine and Pharmacy "Iuliu Hatieganu", 8 Victor Babes Street, 400012 Cluj-Napoca, Romania; motoascairina@gmail.com (I.M.); irsaylaszlo@gmail.com (L.I.)

[5] Faculty of Dental Medicine, "Ovidius" University of Constanta, 7 Ilarie Voronca Street, 900178 Constanta, Romania; crilucpetcu@gmail.com

* Correspondence: iliescumadalina@gmail.com (M.-G.I.); viorela.ciortea@yahoo.com (V.M.C.)

Citation: Stanciu, L.-E.; Iliescu, M.-G.; Oprea, C.; Ionescu, E.-V.; Petcu, A.; Ciortea, V.M.; Petcu, L.C.; Apostol, S.; Nedelcu, A.-D.; Motoașcă, I.; et al. The Impact of Complex Rehabilitation Treatment on Sarcopenia—Pathology with an Endocrine Morphological Substrate and Musculoskeletal Implications. *Medicina* **2023**, *59*, 1238. https://doi.org/10.3390/medicina59071238

Academic Editor: Alberto Di Martino

Received: 27 April 2023
Revised: 17 June 2023
Accepted: 29 June 2023
Published: 2 July 2023

Abstract: The pathogenesis of sarcopenia is multifactorial, including changes in the endocrine system. Easy-to-perform screening tests can guide the diagnosis of sarcopenia and the rehabilitation therapeutic conduct, which can act on many physiopathological links. This study was conducted over a period of 5 months, from April to August 2022, and included 84 patients hospitalized for a period of 10 days in the Balneal and Rehabilitation Sanatorium Techirghiol for complex physiotherapy, which included balneotherapy. In dynamics, both at admission and discharge, specific screening tests for sarcopenia (SARC-F questionnaire, grip strength, testing muscle strength at the level of the quadriceps, sit-to-stand tests (the time required for five consecutive rises and the number of rises performed in 30 s)) and the Visual Analogue Scale (VAS) for pain were performed. The study was conducted according to the norms of deontology and medical ethics. *Results*: A significant proportion of patients had a positive result in at least one of the tests for the screening of sarcopenia syndrome. The most eloquent results were obtained from the statistical analysis of the following parameters evaluated at admission: the SARC-F questionnaire and the sit-to-stand test—the number of rises in 30 s. In terms of dynamics, after performing the complex rehabilitation treatment, the patients recorded improvements in the established screening tests and improvements in pain symptoms evaluated with the help of the VAS. *Conclusions*: Sarcopenia, a pathology developed with aging, is frequently encountered among adults. In the future, it is important to perform screening for sarcopenia in both endocrinology and medical rehabilitation clinics. Good management of sarcopenia can influence therapeutic conduct and can prevent complications, improving the functional capacity and the quality of life of the patients.

Keywords: sarcopenia; endocrine substrate; SARC-F; rehabilitation

1. Introduction

The first hypothesis of the study is that sarcopenia is a frequent but underdiagnosed pathology among adults and the elderly, which is why performing scientifically validated screening tests is very important. The second study hypothesis supports the endocrine physiopathological substrate of sarcopenia and the possibility of obtaining a positive effect of the complex rehabilitation treatment in optimizing the functionality of the endocrine glands and the activity of proinflammatory cytokines.

Sarcopenia syndrome is characterized by the progressive loss of muscle mass, strength and function, starting with the 4th decade of life, in older adults [1]. It is a frequent pathology, and its prevalence increases with the aging of the population [2]. This syndrome represents a health problem that is frequently undiagnosed and has an impact on morbidity, mortality and healthcare expenditure [3]. The causes are multifactorial, involving the hormonal system, endocrinological decline, fatty infiltration, poor nutrition, alcohol consumption, smoking, the activity of the inflammatory pathway, neurological decline, loss of neuromuscular junctions and chronic diseases [1,4]. Sarcopenia is an underestimated and underdiagnosed pathology among geriatric patients, especially in the case of those with a history of stroke or malnutrition [5,6].

The consequences of sarcopenia are fatigue, falls, fractures, loss of function, frailty, loss of independence and disability [1]. For the good management of the progressive loss of muscle mass, strength and function, personalized and effective strategies for the prevention, diagnosis from an early stage and treatment of sarcopenia must be adopted [2]. To improve the functional status, the management of sarcopenia focuses on increasing physical activity and ensuring adequate nutrition, even from adulthood [1,7].

Since 2016, sarcopenia has been recognized as a disease entity under the following diagnostic code: ICD_10_M62.8 [8,9]. Measuring muscle mass is a cornerstone in establishing the diagnosis. This measurement can be performed by dual-energy X-ray absorptiometry (DXA), magnetic resonance imaging (MRI) and computed tomography (CT) [10]. Additionally, morphological estimates of skeletal muscle can be based on anthropometric measurements or, in a more modern approach, on ultrasound measurements [4,8].

These methods can be expensive and time-consuming and may require experienced medical personnel, which is why screening tests must be performed beforehand. Some useful screening tools will be described in the present study.

The pathogenesis of sarcopenia can be supported by the increased levels of tumor necrosis factor alpha (TNF-α) that occur with aging, being associated with the loss of muscle fibers and apoptosis [11]. Also, high concentrations of interleukin 6 (IL-6) may play an important role in modulating the inflammatory pathway during skeletal muscle loss [12].

The changes in the endocrine system that occur with age could be trigger factors for sarcopenia because the endocrine system has complex interactions with skeletal muscles, contributing to muscle development and the modulation of muscle strength [13]. The decline of hormones that maintain muscle (insulin-like growth factor (IGF-1), dehydroepiandrosterone sulfate (DHEA-S), testosterone, estrogen) can contribute to the appearance of sarcopenia [13]. The hypothalamic–pituitary–adrenal axis becomes dysfunctional with age, causing the adrenal cortex to release more glucocorticoids that induce a decrease in protein synthesis by affecting myostatin and IGF-1 [13]. Vitamin D deficiency can increase the risk of sarcopenia. Several hypotheses claim that supplementation with vitamin D can be effective in reaching and treating this syndrome.

A study published in 2011 demonstrates that sapropelic mud therapy influences the serum level of proinflammatory cytokines. The patients selected in the study with increased values of TNF-α, IL-1β and IL-6 recorded normal values at discharge, after ten days of treatment [10].

The findings show that the clinical benefits of balneotherapy in musculoskeletal disease patients may be mediated by an anti-inflammatory effect. A decrease in circulating IL-6 levels was noticed, as were beneficial improvements in pain decrease and patient functionality [14].

Currently, due to the lack of knowledge and due to the multifactorial genesis, the approach for a patient with sarcopenia is not based on basic etiopathogenesis, and the manifestations of sarcopenia are frequently masked by the multiple comorbidities that elderly patients may have [13,15].

The main objectives of this study were to demonstrate the importance of screening tests and to evaluate the impact of multimodal rehabilitation treatment on patients with sarcopenia, a pathology with an endocrine morphological substrate.

2. Methods

2.1. Study Design

This was a prospective, longitudinal, before–after study conducted over a period of 5 months, from April to August 2022, in a representative unit in the field of medical rehabilitation in Romania.

2.2. Subjects

The study included 84 patients, aged between 50 and 79 years, hospitalized for 10 days in Techirghiol Balneal and Rehabilitation Sanatorium for balneary–physical–kinetic treatment.

The inclusion criteria for the patients were as follows:

- Hospitalized in Balneal and Rehabilitation Sanatorium Techirghiol for musculoskeletal disorders;
- Aged between 50 and 79 years;
- Conscious, cooperative;
- The possibility of performing all the pre-established screening tests.

The exclusion criteria for the patients were as follows:

- Outpatient treatment for musculoskeletal disorders;
- Age outside the selected range;
- Uncooperative patients;
- Comorbidities associated with the impossibility of performing all the pre-established screening tests.

2.3. Intervention

The patients were evaluated, by screening tests, in dynamics, both at admission and discharge to observe the impact of complex rehabilitation treatment on muscle mass, strength and function in the context of sarcopenia.

The rehabilitation treatment performed by the patients was standard, according to the existing treatment protocols at the level of the medical unit and the national and international balneological approach. The interventional treatment included 10 daily sessions of hydrokinetotherapy and hydrothermotherapy using specific natural environmental factors, namely sapropelic mud and salt water from Techirghiol Lake, as well as artificial physical agents such as electrotherapy, massage and of course exercise therapy.

The baths in the salty water of Lake Techirghiol (at 35 °C) and mud baths (at 38 °C) were alternated daily, for a period of 20 min. The hydrokinetotherapy was performed on the recommendation of the rehabilitation doctor and under the guidance and supervision of the rehabilitation assistant. At the recommendation of the rehabilitation doctor, each patient performed, depending on the associated musculoskeletal pathology, three electrotherapy procedures daily. Massage and exercise therapy were performed on the doctor's recommendation by the rehabilitation assistant, daily. For each patient included in the study, the same therapeutic objectives were respected to obtain a similar functional impact.

2.4. Measures

The SARC-F questionnaire, specific functional tests for sarcopenia and the Visual Analogue Scale were performed and analyzed. The study protocol is presented in Figure 1.

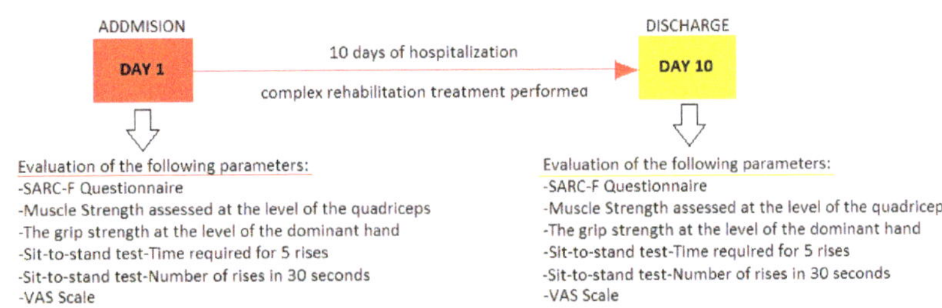

Figure 1. Protocol timeline of the study.

The SARC-F questionnaire is a screening tool, easy to use in daily practice, made up of 5 questions (strength, walking assistance, lifting from a chair, climbing stairs, falling) [3]. Both at admission and at discharge, on day 1 and day 10, the patients completed the questionnaire and evaluated each question with 0–2 points. A score greater than or equal to 4 gives a positive result in the screening test performed with the help of this tool [3].

Other sarcopenia screening tools that evaluate muscle strength and performance are represented by grip strength (assessed with a dynamometer), testing muscle strength at the level of the quadriceps, sit-to-stand tests (the time required for five consecutive rises and the number of rises performed in 30 s).

Both at admission and at discharge, a subjective evaluation method that depends on the evaluator was used to evaluate quadriceps strength. On the first day and last day of hospitalization, the patient performed a calf extension while sitting, with the knee flexed at 90° and the leg outside the table, while the doctor put resistance on the lower third of the calf. This test was performed to assess if the patient had low muscle strength at this level.

The grip strength was evaluated with the help of a digital hand dynamometer at the level of the dominant hand both at admission (day 1) and at discharge (day 10). Suggestive values for sarcopenia depend on the sex of the patient. Positive screening for sarcopenia was considered <16 kg for women and <27 kg for men [7].

Sit-to-stand tests were performed both at admission (first day) and at discharge (last day). The patients were asked to stand up and sit on a chair with feet about shoulder width apart and the arms crossed at the wrist and held against the chest. The time taken to rise from the chair 5 times was recorded. A value greater than 15 s provided a positive screening for sarcopenia [7]. The second test consisted in counting the rises performed in 30 s. The positive screening depended on the patient's sex and age (Table 1).

Table 1. Positive screening for sarcopenia by using sit-to-stand test—the number of rises in 30 s.

Age	Male	Female
50–54 years old	<16 rises	<14 rises
55–59 years old	<15 rises	<13 rises
60–64 years old	<14 rises	<12 rises
65–69 years old	<12 rises	<11 rises
70–74 years old	<12 rises	<10 rises
75–79 years old	<11 rises	<10 rises

In order to observe how many patients in the investigated group had a positive screening for sarcopenia, all five tests were performed.

The Visual Analogue Scale represents a subjective method for chronic and acute pain assessment [16]. This scale, which is easy and quick to perform, was applied to all the patients included in the study. To evaluate if the rehabilitation treatment improved the pain syndrome, patients were verbally asked by the clinician where the pain is on a scale from 0 (no pain) to 10 (worst pain imaginable), both at admission and at discharge.

2.5. Data Analysis

Each screening test performed at admission was analyzed as follows:

- In relation to certain fixed parameters (sex, environment, age, weight status);
- In relation to the Visual Analogue Scale (VAS) at admission/discharge;
- In relation to the other screening tests performed at admission;
- In relation to the result obtained in the same screening test at discharge (dynamic evaluation by performing a statistical association).

Statistical correlations were determined to identify the parameters of the screening tests that correlate best, as well as which screening test is more suggestive of sarcopenia.

The statistical analysis was performed using IBM SPSS statistics software version 23. The following tests were used in the statistical analysis of the obtained data: Wilcoxon signed-rank test and Z test for associations, chi-square test and Spearman's rank-order for correlations.

The study protocol was designed in accordance with the ethical guidelines of the Declaration of Human Rights and was approved by the Techirghiol Balneal and Rehabilitation Sanatorium Ethics Committee.

3. Results

Of the 84 analyzed patients, 61% were women, 82.1% were from urban areas, and 47.6% were aged between 50 and 59 years. It was found that 85.71% of the investigated patients were positive in at least one screening test performed for sarcopenia. Statistically significant associations and correlations were observed between the performed screening tests and between them and the rest of the evaluated parameters.

3.1. Results Obtained According to Each Screening Test

(1) SARC-F questionnaire is suggestive of sarcopenia when a value higher than 4 is obtained. A percentage of 27.38% of patients had a positive screening for sarcopenia through the SARC-F questionnaire at admission (Figure 2).

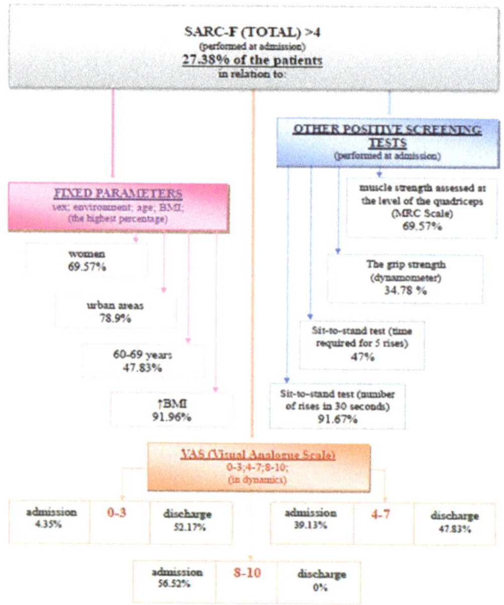

Figure 2. Schematic description of the SARC-F questionnaire (total), performed at admission.

The evaluation of the SARC-F questionnaire at discharge, after the complex rehabilitation treatment was performed, identified 13.10% of the total analyzed patients with a positive screening for sarcopenia. The association between SARC-F (total) at admission and discharge using the Wilcoxon signed-rank test and the Z test was statistically relevant (Figure 3).

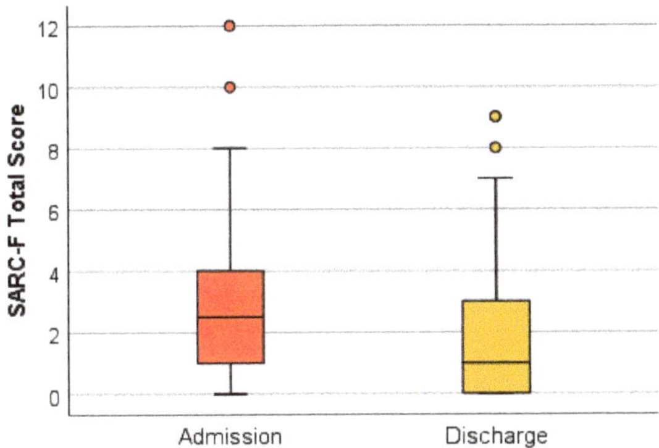

Figure 3. SARC-F (total) at admission and discharge.

The Wilcoxon signed-rank test determined that there was a statistically significant median decrease in SARC-F (total) score at admission (2.50) compared to SARC-F (total) score at discharge (1.00), $Z = -5.456$, $p < 0.0001$.

(2) The testing of muscle strength at the level of the quadriceps was performed according to the MRC scale (Medical Research Council Scale for Power of Muscle). A percentage of 40.48% of the evaluated patients recorded decreases in muscle strength (Figure 4).

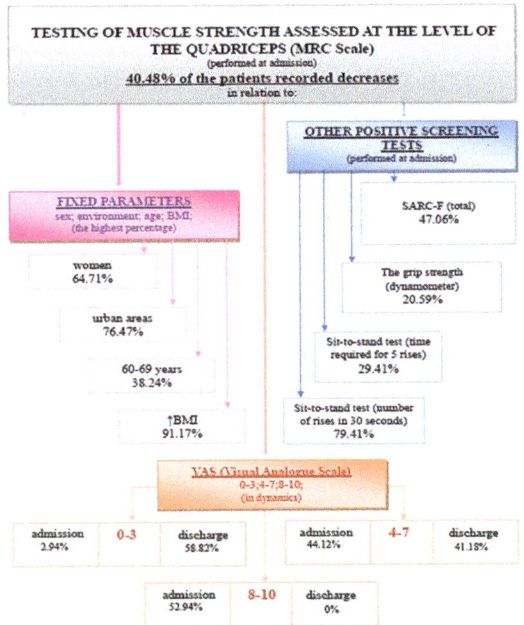

Figure 4. Schematic description of the muscle strength assessed at the level of the quadriceps (MRC Scale), performed at admission.

The evaluation of muscle strength at the quadriceps level at discharge confirms that 30.95% of the total 84 patients registered a decrease in muscle strength. The association between quadriceps muscle strength at admission and discharge using the Wilcoxon signed-rank test and Z test has no statistical significance (Figure 5).

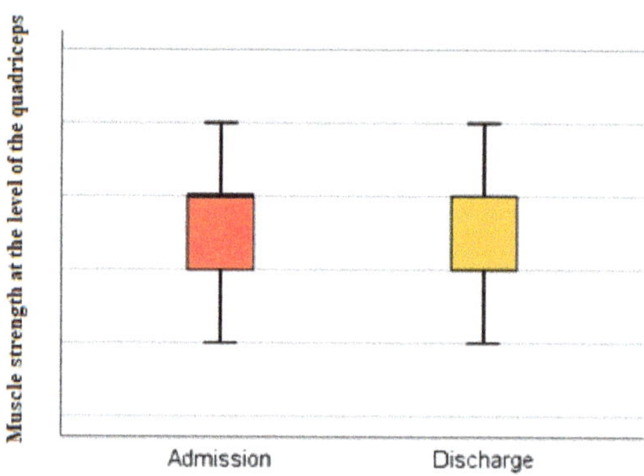

Figure 5. Muscle strength at the level of the quadriceps at admission and discharge.

The Wilcoxon signed-rank test showed that at discharge, the treatment did not elicit a statistically significant change in testing muscle strength at the level of the quadriceps ($Z = -1.633$, $p = 0.102$). Indeed, the median score rating was the same at admission and at discharge.

(3) The evaluation of the grip strength at the level of the dominant hand was performed with the help of a dynamometer. It is observed that 14.29% of the patients registered values compatible with sarcopenia at admission (Figure 6).

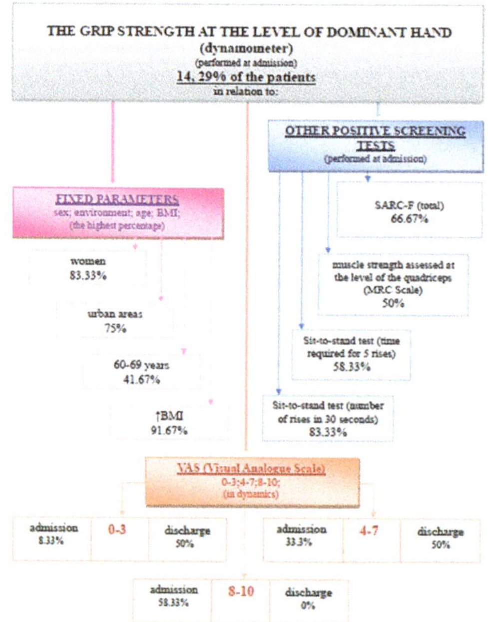

Figure 6. Schematic description of the grip strength at the level of dominant hand, performed at admission.

At discharge, 11.9% of patients had a positive screening for sarcopenia through this evaluation method. The association of the grip strength at admission and at discharge using the Wilcoxon signed-rank test and the Z test was statistically significant (Figure 7).

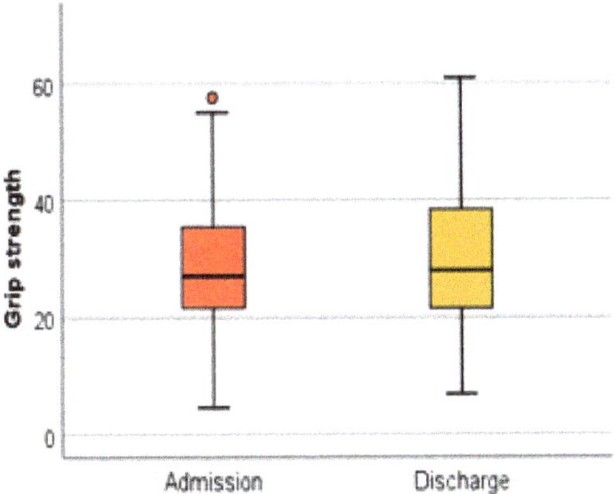

Figure 7. Grip strength at admission and discharge.

The Wilcoxon signed-rank test determined that there was a statistically significant median increase in grip strength (kg) at admission compared to grip strength (kg) at discharge, $Z = -2.958$, $p < 0.0001$.

(4) Performing the sit-to-stand test (the time required for five consecutive rises) recorded a positive screening for sarcopenia (>15 s) in 25% of patients (Figure 8).

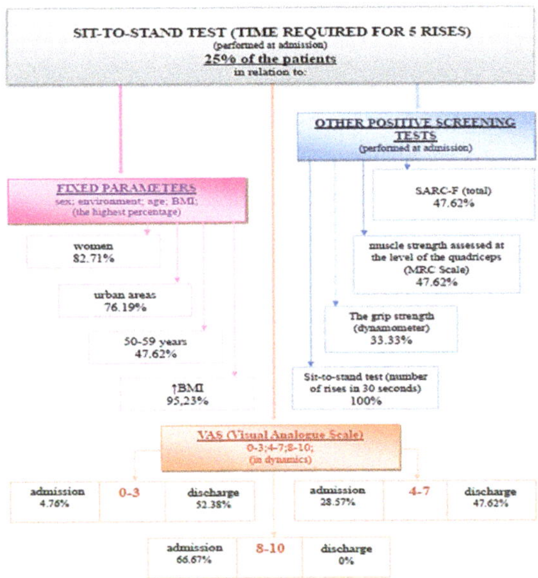

Figure 8. Schematic description of the sit-to-stand test (time required for 5 rises), performed at admission.

Performing the sit-to-stand test (the time required for five consecutive rises) at discharge recorded values for sarcopenia in 14% of cases. The association of this parameter at admission and at discharge using the Wilcoxon signed-rank test and Z test had statistical significance (Figure 9).

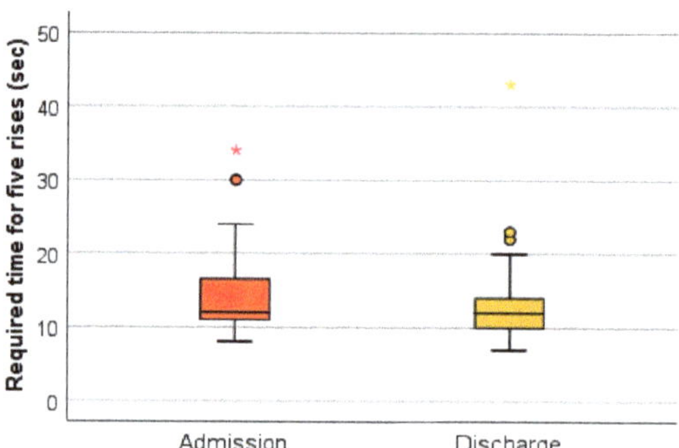

Figure 9. The time required for five consecutive rises at admission and discharge. */★—cohort.

The Wilcoxon signed-rank test determined that there was a statistically significant median decrease in the time required for five consecutive rises at admission (12.65) compared to the time required for five consecutive rises at discharge (12.00), $Z = -4.468$, $p < 0.0001$.

(5) The interpretation of the results of the sit-to-stand test (the number of rises in 30 s) depends on the patient's sex and age (Table 1).

This test used as a screening method for sarcopenia was positive for 76.19% of the 84 evaluated patients (Figure 10).

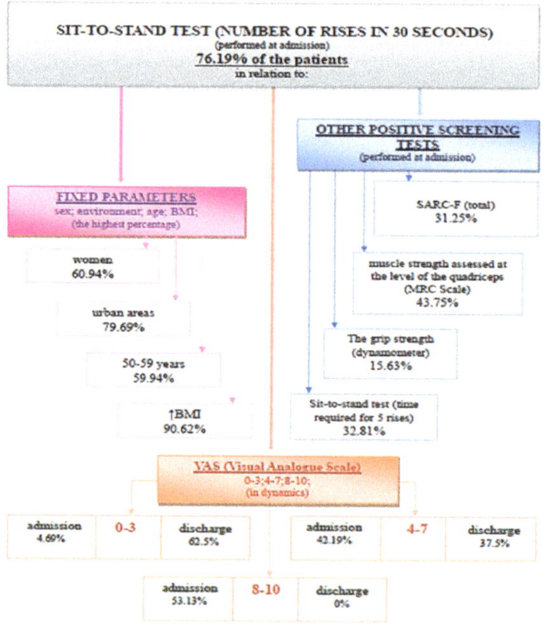

Figure 10. Schematic description of the sit-to-stand test (number of rises in 30 s), performed at admission.

At discharge, 57.14% of patients had a positive screening for sarcopenia through the sit-to-stand test (the number of rises in 30 s). The association of this parameter at admission and at discharge using the Wilcoxon signed-rank test and Z test had statistical significance (Figure 11).

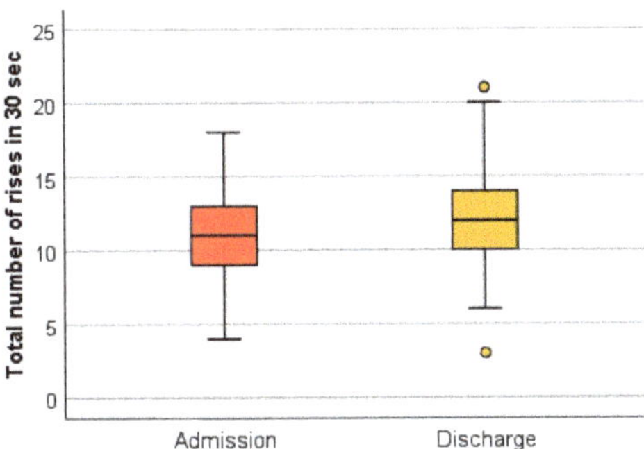

Figure 11. The sit-to-stand test (number of rises in 30 s) at admission and discharge.

The Wilcoxon signed-rank test determined that there was a statistically significant median increase in the sit-to-stand test (the number of rises in 30 s) at admission (11.00) compared to the sit-to-stand test (the number of rises in 30 s) at discharge (12.00), $Z = -5.405, p < 0.0001$.

3.2. VAS Scores in Dynamics (Statistical Association)

The VAS scores obtained at admission and discharge were associated and evaluated using the Wilcoxon signed-rank test and the Z test. The results obtained showed that the algo-dysfunctional syndrome evaluated with the help of the Visual Analogue Scale (VAS), present at admission, improved during hospitalization, with significant relief of the pain (Figure 12).

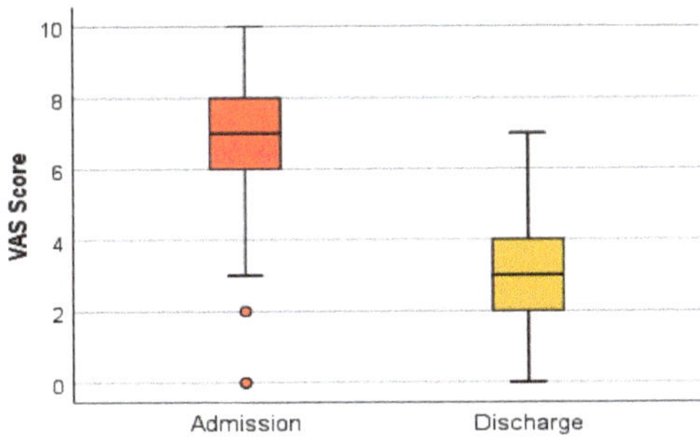

Figure 12. VAS score at admission and discharge.

The Wilcoxon signed-rank test determined that there was a statistically significant median decrease in VAS score at admission (7.00) compared to VAS score at discharge (3.00), $Z = -7.960, p < 0.0001$.

3.3. Relevant Statistical Correlations Identified between the Associated Parameters

(1) By performing chi-square tests, it was observed that there was a statistically significant association between sex and grip strength (kg) at admission ($p = 0.000 < 0.05$), sex and the sit-to-stand test (the time required for five consecutive rises (seconds)) at admission ($p = 0.011 < 0.05$) and sex and the sit-to-stand test (the number of rises in 30 s) at admission ($p = 0.016 < 0.05$) (Tables 2–4).

Table 2. Sex grip strength dominant hand (kg) at admission.

			Grip Strength Dominant Hand (kg) at Admission			Total
			0–20 kg	20.1–35.5 kg	>35.5 kg	
Sex	Female	Count	12	38	1	51
		% of Total	14.3%	45.2%	1.2%	60.7%
	Male	Count	1	13	19	33
		% of Total	1.2%	15.5%	22.6%	39.3%
Total		Count	13	51	20	84
		% of Total	15.5%	60.7%	23.8%	100.0%
Chi-Square Tests			Value	df		p
Pearson Chi-Square			35.537	2		0.000

Table 3. Sex required time for five rises at admission (s).

			Required Time for Five Rises at Admission (s)				Total
			0–9 s	10–15 s	16–20 s	>20 s	
Sex	Female	Count	1	31	13	6	51
		% of Total	1.2%	36.9%	15.5%	7.1%	60.7%
	Male	Count	4	26	3	0	33
		% of Total	4.8%	31.0%	3.6%	0.0%	39.3%
Total		Count	5	57	16	6	84
		% of Total	6.0%	67.9%	19.0%	7.1%	100.0%
Chi-Square Tests			Value	df		p	
Pearson Chi-Square			11.143	3		0.011	

Table 4. Sex total number of rises in 30 s at admission.

			Total Number of Rises in 30 s at Admission				Total
			<5 Rises	5–10 Rises	11–15 Rises	>15 Rises	
Sex	Female	Count	1	26	23	1	51
		% of Total	1.2%	31.0%	27.4%	1.2%	60.7%
	Male	Count	0	6	26	1	33
		% of Total	0.0%	7.1%	31.0%	1.2%	39.3%
Total		Count	1	32	49	2	84
		% of Total	1.2%	38.1%	58.3%	2.4%	100.0%
Chi-Square Tests			Value		df		p
Pearson Chi-Square			10.299		3		0.016

(2) By performing chi-square tests, it was observed that there was a statistically significant association between age and quadriceps muscle strength at discharge (p = 0.024 < 0.05) and between age and the sit-to-stand test (the number of rises in 30 s) at discharge (p = 0.002 < 0.05) (Tables 5 and 6).

(3) By performing chi-square tests, relevant statistical correlations were identified between weight status and the sit-to-stand test (the number of rises in 30 s) at admission (p = 0.000 < 0.05).

(4) By performing chi-square tests and Spearman's rank-order correlation, relevant statistical correlations were identified between the SARC-F (total) and the other established tests of screening.

Table 5. Age quadriceps muscle strength at discharge.

| | | | Quadriceps Muscle Strength at Discharge | | | | Total |
			5	−5	4	−4	
Age	50–59	Count	32	8	0	0	40
		% of Total	38.1%	9.5%	0.0%	0.0%	47.6%
	60–69	Count	25	3	0	0	28
		% of Total	29.8%	3.6%	0.0%	0.0%	33.3%
	70–79	Count	10	3	2	1	16
		% of Total	11.9%	3.6%	2.4%	1.2%	19.0%
Total		Count	67	14	2	1	84
		% of Total	79.8%	16.7%	2.4%	1.2%	100.0%

Chi-Square Tests		
	Value	df	p
Pearson Chi-Square	14.570	6	0.024

Table 6. Age total number of rises in 30 s at discharge.

| | | | Total Number of Rises in 30 s at Discharge | | | | Total |
			<5 Rises	5–10 Rises	11–15 Rises	>15 Rises	
Age	50–59	Count	1	9	28	2	40
		% of Total	1.2%	10.7%	33.3%	2.4%	47.6%
	60–69	Count	0	6	13	9	28
		% of Total	0.0%	7.1%	15.5%	10.7%	33.3%
	70–79	Count	0	10	5	1	16
		% of Total	0.0%	11.9%	6.0%	1.2%	19.0%
Total		Count	1	25	46	12	84
		% of Total	1.2%	29.8%	54.8%	14.3%	100.0%

Chi-Square Tests		
	Value	df	p
Pearson Chi-Square	21.278	6	0.002

Statistically significant associations were found between SARC-F (total) at admission and quadriceps muscle strength at admission ($p = 0.002 < 0.05$), SARC-F (total) at admission and grip strength (kg) at admission ($p = 0.000 < 0.05$), SARC-F (total) at admission and the sit-to-stand test (the time required for five consecutive rises (seconds)) at admission ($p = 0.000 < 0.05$) and SARC-F (total) at admission and the sit-to-stand test (the number of rises in 30 s) at admission ($p = 0.013 < 0.05$) (Tables 7–10).

Table 7. SARC-F total at admission quadriceps muscle strength at admission.

| | | | Quadriceps Muscle Strength at Admission | | | | Total |
			5	−5	4	−4	
SARC-F Total at admission	0–3	Count	51	6	5	0	62
		% of Total	60.7%	7.1%	6.0%	0.0%	73.8%
	4–7	Count	8	5	3	1	17
		% of Total	9.5%	6.0%	3.6%	1.2%	20.2%
	8–12	Count	1	1	2	1	5
		% of Total	1.2%	1.2%	2.4%	1.2%	6.0%
Total		Count	60	12	10	2	84
		% of Total	71.4%	14.3%	11.9%	2.4%	100.0%

Chi-Square Tests		
	Value	df	p
Pearson Chi-Square	21.466	6	0.002

Table 8. SARC-F total at admission grip strength dominant hand (kg) at admission.

			Grip Strength Dominant Hand (kg) at Admission			Total
			0–20 kg	**20.1–35.5 kg**	**>35.5 kg**	
SARC-F Total at admission	0–3	Count	3	43	16	62
		% of Total	3.6%	51.2%	19.0%	73.8%
	4–7	Count	5	8	4	17
		% of Total	6.0%	9.5%	4.8%	20.2%
	8–12	Count	5	0	0	5
		% of Total	6.0%	0.0%	0.0%	6.0%
Total		Count	13	51	20	84
		% of Total	15.5%	60.7%	23.8%	100.0%
Chi-Square Tests						
		Value	df		p	
Pearson Chi-Square		35.363	4		0.000	

Table 9. SARC-F total at admission required time for five rises at admission (s).

			Required Time for Five Rises at Admission (s)				Total
			0–9 s	**10–15 s**	**16–20 s**	**>20 s**	
SARC-F Total at admission	0–3	Count	4	47	11	0	62
		% of Total	4.8%	56.0%	13.1%	0.0%	73.8%
	4–7	Count	1	10	3	3	17
		% of Total	1.2%	11.9%	3.6%	3.6%	20.2%
	8–12	Count	0	0	2	3	5
		% of Total	0.0%	0.0%	2.4%	3.6%	6.0%
Total		Count	5	57	16	6	84
		% of Total	6.0%	67.9%	19.0%	7.1%	100.0%
Chi-Square Tests							
		Value	df		p		
Pearson Chi-Square		32.336	6		0.000		

Table 10. SARC-F total at admission total number of rises in 30 s at admission.

			Total Number of Rises in 30 s at Admission				Total
			<5 Rises	**5–10 Rises**	**11–15 Rises**	**>15 Rises**	
SARC-F Total at admission	0–3	Count	0	19	42	1	62
		% of Total	0.0%	22.6%	50.0%	1.2%	73.8%
	4–7	Count	1	8	7	1	17
		% of Total	1.2%	9.5%	8.3%	1.2%	20.2%
	8–12	Count	0	5	0	0	5
		% of Total	0.0%	6.0%	0.0%	0.0%	6.0%
Total		Count	1	32	49	2	84
		% of Total	1.2%	38.1%	58.3%	2.4%	100.0%
Chi-Square Tests							
		Value	df		p		
Pearson Chi-Square		16.096	6		0.013		

A Spearman's rank-order correlation was run to determine the relationship between SARC-F (total) at admission and grip strength (kg) at admission/the sit-to-stand test (the time required for five consecutive rises (seconds)) at admission / the sit-to-stand test (the number of rises in 30 s) at admission.

There was a moderate, negative correlation between SARC-F (total) at admission and grip strength (kg) at admission, which was statistically significant ($p < 0.001$); a moderate, positive correlation between SARC-F (total) at admission and the sit-to-stand test (the time required for five consecutive rises (seconds)) at admission, which also was statistically significant ($p < 0.001$); and a moderate, negative correlation between SARC-F (total) at admission and the sit-to-stand test (the number of rises in 30 s) at admission, which also was statistically significant ($p < 0.001$) (Table 11).

Table 11. Spearman's rank-order correlation between SARC-F (total) at admission and other parameters.

			Grip Strength Right Hand (kg) at Admission	Required Time for Five Rises at Admission (s)	Total Number of Rises in 30 s at Admission
		Nonparametric Correlations			
		Correlations			
Spearman's rho	SARC-F Total at admission	Correlation Coefficient	−0.409	0.496	−0.484
		Sig. (2-tailed)	0.000	0.000	0.000
		N	84	84	84

There was a statistically significant association between SARC-F (total) at discharge and quadriceps muscle strength at discharge ($p = 0.002 < 0.05$), SARC-F (total) at discharge and grip strength (kg) at discharge ($p = 0.001 < 0.05$), SARC-F (total) at discharge and the sit-to-stand test (the time required for five consecutive rises (seconds)) at discharge ($p = 0.000 < 0.05$) and between SARC-F (total) at discharge and the sit-to-stand test (the number of rises in 30 s) at discharge ($p = 0.001 < 0.05$) (Tables 12–15).

Table 12. SARC-F total at discharge quadriceps muscle strength at discharge.

			Quadriceps Muscle Strength at Discharge				Total
			5	−5	4	−4	
SARC-F Total at discharge	0–3	Count	60	8	2	0	70
		% of Total	71.4%	9.5%	2.4%	0.0%	83.3%
	4–7	Count	4	6	0	1	11
		% of Total	4.8%	7.1%	0.0%	1.2%	13.1%
	8–12	Count	3	0	0	0	3
		% of Total	3.6%	0.0%	0.0%	0.0%	3.6%
Total		Count	67	14	2	1	84
		% of Total	79.8%	16.7%	2.4%	1.2%	100.0%

	Chi-Square Tests		
	Value	df	*p*
Pearson Chi-Square	21.221	6	0.002

Table 13. SARC-F total at discharge grip strength dominant hand (kg) at discharge.

			Grip Strength Dominant Hand (kg) at Discharge			Total
			0–20 kg	20.1–35.5 kg	>35.5 kg	
SARC-F Total at discharge	0–3	Count	8	40	22	70
		% of Total	9.5%	47.6%	26.2%	83.3%
	4–7	Count	4	6	1	11
		% of Total	4.8%	7.1%	1.2%	13.1%
	8–12	Count	3	0	0	3
		% of Total	3.6%	0.0%	0.0%	3.6%
Total		Count	15	46	23	84
		% of Total	17.9%	54.8%	27.4%	100.0%

	Chi-Square Tests		
	Value	df	*p*
Pearson Chi-Square	19.365	4	0.001

Table 14. SARC-F total at discharge required time for five rises at discharge (s).

| | | | Required Time for Five Rises at Discharge (s) | | | | Total |
			0–9 s	10–15 s	16–20 s	>20 s	
SARC-F Total at discharge	0–3	Count	18	47	4	1	70
		% of Total	21.4%	56.0%	4.8%	1.2%	83.3%
	4–7	Count	0	6	3	2	11
		% of Total	0.0%	7.1%	3.6%	2.4%	13.1%
	8–12	Count	0	1	0	2	3
		% of Total	0.0%	1.2%	0.0%	2.4%	3.6%
Total		Count	18	54	7	5	84
		% of Total	21.4%	64.3%	8.3%	6.0%	100.0%

Chi-Square Tests	Value	df	p
Pearson Chi-Square	33.608	6	0.000

Table 15. SARC-F total at discharge total number of rises in 30 s at discharge.

| | | | Total Number of Rises in 30 s at Discharge | | | | Total |
			<5 Rises	5–10 Rises	11–15 Rises	>15 Rises	
SARC-F Total at discharge	0–3	Count	0	15	43	12	70
		% of Total	0.0%	17.9%	51.2%	14.3%	83.3%
	4–7	Count	1	7	3	0	11
		% of Total	1.2%	8.3%	3.6%	0.0%	13.1%
	8–12	Count	0	3	0	0	3
		% of Total	0.0%	3.6%	0.0%	0.0%	3.6%
Total		Count	1	25	46	12	84
		% of Total	1.2%	29.8%	54.8%	14.3%	100.0%

Chi-Square Tests	Value	df	p
Pearson Chi-Square	23.612	6	0.001

A Spearman's rank-order correlation was run to determine the relationship between SARC-F (total) at discharge and grip strength (kg) at discharge/the sit-to-stand test (the time required for five consecutive rises (seconds)) at discharge/the sit-to-stand test (the number of rises in 30 s) at discharge.

There was a moderate, negative correlation between SARC-F (total) at discharge and grip strength (kg) at discharge, which was statistically significant ($p < 0.001$); a weak, positive correlation between SARC-F (total) at discharge and the sit-to-stand test (the time required for five consecutive rises (seconds)) at discharge, which also was statistically significant ($p < 0.001$); and a moderate, negative correlation between SARC-F (total) at discharge and the sit-to-stand test (the number of rises in 30 s) at discharge, which also was statistically significant ($p < 0.001$) (Table 16).

Table 16. Spearman's rank-order correlation between SARC-F (total) at discharge and other parameters.

			Grip Strength Dominant Hand (kg) at Discharge	Required Time for Five Rises at Discharge (s)	Total Number of Rises in 30 s at Discharge
Spearman's rho	SARC-F Total at discharge	Correlation Coefficient	−0.494	0.378	−0.479
		Sig. (2-tailed)	0.000	0.000	0.000
		N	84	84	84

4. Discussion

It is important to have a screening strategy that allows the detection of sarcopenia in the early stages because it is extremely difficult to regain the already lost skeletal muscle mass.

An ideal screening test must be useful, valid, reliable and also cost-effective. Screening strategies can be carried out through several tools, and it is necessary to identify which is the most effective way to detect sarcopenia and to choose some methods optimal for the prevention and treatment of this pathology [3].

The evaluation and diagnosis of sarcopenia can represent a new objective in defining the incidence and prevalence of this pathology, which is why it is necessary to establish a solid and comprehensive algorithm. Refinement of the algorithm and its use in current practice can represent a tool for monitoring progress [4].

The results of the study demonstrated that our objectives were achieved, the screening tests proved their usefulness in identifying sarcopenia, and rehabilitation treatment can have an effect on sarcopenia through the possible effects on the endocrine system.

To our knowledge, this is the only study using balneotherapy for sarcopenic patients. Other studies which applied physiotherapy as a therapeutic intervention had good results for exercise therapy, while other methods were not largely studied [3,15,17–20].

In this study, it was observed that although 85.71% of the patients were positive in at least one screening test, none were previously investigated for sarcopenia, and also none of them had a diagnosis of sarcopenia in their pathological history. A percentage of 27.38% of the evaluated patients had a positive screening for sarcopenia through the SARC-F questionnaire at admission. A percentage of 40.48% of the patients recorded decreases in the testing of muscle strength at the level of the quadriceps at admission. It is observed that 14.29% of the patients registered values compatible with sarcopenia in the evaluation of grip strength at the level of the dominant hand. The time required for five consecutive rises in the sit-to-stand test recorded a positive screening for sarcopenia for 25% of the patients. The number of rises in 30 s in the sit-to-stand test as a screening method for sarcopenia was positive for 76.19% of the evaluated patients.

The dynamic assessment of screening tests, both at admission and discharge, suggests that complex rehabilitation treatment can have an impact on sarcopenia. The decrease in the SARC-F (total) score at discharge indicates an improvement in functionality. This may be due to the decrease in the algo-functional syndrome and the increase in general muscle strength after performing the complex rehabilitation treatment. The testing of quadriceps muscle strength did not register considerable changes because this method is subjective and the improvements are difficult to quantify. The muscle strength measured with the dynamometer recorded a statistically significant increase. This method of evaluation uses a standard technique, devoid of subjectivity. The sit-to-stand tests (the time required for five consecutive rises and the number of rises in 30 s) were shortened at discharge, which indicates an increase in muscle strength and resistance following the complex treatment performed during hospitalization. These improvements can be due to both the hormonal implications (that complex rehabilitation treatment has through the use of natural therapeutic factors) and kinesiotherapy.

It was observed that evaluation of the Visual Analogue Scale (VAS) at discharge recorded improvements, with significant relief of the pain. Protocols that combine specific rehabilitation treatments with balneotherapy, including hydrotherapy, mud baths and exercise in hot thermal mud, could reduce pain and disability in patients identified as responders [21].

Relevant statistical correlations were identified between the associated parameters. The general muscle strength evaluated both at the level of the upper limbs and the lower limbs is related to sex. The endocrine system plays an important role, making the connection between the level of testosterone and estrogen in both sexes and the decline of muscle strength. There is a close relationship between age and muscle strength in the lower limbs. At discharge, the patients in the lower age ranges recorded improvements in the functional tests. This fact can be explained by hormonal status in relation to the patient's age. Most of the investigated patients have an increased body mass index. The obtained results support the concept of sarcopenic obesity from the literature [22–26]. Thus, the excess representation of adipose tissue can influence functionality.

It was found that the SARC-F questionnaire correlates, both at admission and discharge, with all other screening tests. Considering this, in daily practice, only the SARC-F questionnaire and the number of rises performed in 30 s in the sit-to-stand test can be used for the screening of sarcopenia.

The tests indicated for the diagnosis of sarcopenia can be performed for all cooperative, conscious patients who have multiple associated diseases. Likewise, rehabilitation treatments that use artificial and natural physical agents can be recommended to patients with kidney, liver or heart damage, without determining the side effects of drug treatments [7–10,13,20,21] The physical factors recommended in physical and rehabilitation medicine are considered environmentally friendly alternatives compared to drugs for the treatment of various pathologies [7,10].

4.1. Limitations of the Study

Because the screening does not always confirm the diagnosis and more advanced diagnostic methods are needed, no control group was established. The lack of a control group can be considered a limitation of the study, along with the high costs that advanced sarcopenia diagnostic methods can have. Another limitation of the study is that no studies were found in the literature about sarcopenia and complex rehabilitation treatment that includes balneotherapy, which could have an impact on the endocrine system.

4.2. Applications and Suggestions for Future Research

In the future, performing screening tests in endocrinology and medical rehabilitation clinics can guide the establishment of diagnosis and treatment of sarcopenia. Even though there have been efforts to standardize a blood test kit for sarcopenia, there is currently no such kit in use in clinical settings [18]. For future studies, a larger number of patients, a control group and more diagnostic methods will be used.

5. Conclusions

The interpretation of our findings demonstrates that the results of the study support the hypothesis, that the screening tests, which are easy to perform routinely, prove their usefulness in the diagnosis of sarcopenia and that multimodal rehabilitation treatment has an effect on sarcopenia, improving the results of the screening tests.

The link that sarcopenia creates between endocrinology and medical rehabilitation is based on the hormonal changes that balneal treatment with natural environmental factors and kinesiotherapy can induce. The screening tests—SARC-F questionnaire, grip strength (assessed with a dynamometer), testing muscle strength at the level of the quadriceps, and sit-to-stand tests (the time required for five consecutive rises and the number of rises performed in 30 s)—are easy to perform in a short time, are reliable and have minimal costs. The main parameters analyzed at admission with statistically significant results were the SARC-F questionnaire and the number of rises performed in 30 s in the sit-to-stand test, which confirms the importance of screening by these methods in the evaluation of sarcopenia. Positive statistical variations following the therapeutic intervention were obtained for all the parameters included in the research, except the muscle strength at the level of the quadriceps, which underlines the importance of rehabilitation treatment in sarcopenia.

Screening can be easily performed in both endocrinology and medical rehabilitation clinics, and it can guide the conduct of diagnosis and treatment of sarcopenia (food supplements/hormonal supplements/increasing muscle strength with kinesiotherapy/balneal treatment with natural environmental factors). Sarcopenia has an endocrine morphological substrate with multiple functional implications. The approach to this pathology must be multidisciplinary.

Author Contributions: Conceptualization, L.-E.S., M.-G.I., V.M.C. and L.I.; methodology, L.-E.S., M.-G.I., I.M. and A.P.; software validation, C.O., E.-V.I. and L.C.P.; formal analysis, A.P., L.C.P., S.A. and I.M.; investigation, S.A., A.-D.N., I.M., L.C.P., C.O. and E.-V.I.; resources, L.-E.S., V.M.C., A.-D.N. and S.A.; data curation, C.O., A.P. and E.-V.I.; writing—original draft preparation, L.-E.S., A.P., L.I., I.M. and L.C.P.; writing—review and editing, L.-E.S., M.-G.I., E.-V.I., A.-D.N. and V.M.C.; visualization, M.-G.I., V.M.C. and L.I.; supervision, M.-G.I., V.M.C. and L.I.; project administration, L.-E.S., V.M.C., M.-G.I. and L.I. All authors have read and agreed to the published version of the manuscript.

Funding: This research received no external funding.

Institutional Review Board Statement: The study was conducted in accordance with the Declaration of Helsinki and approved by the Techirghiol Balneal and Rehabilitation Sanatorium Ethics Committee (5491/14 April 2022).

Informed Consent Statement: Informed consent was obtained from all subjects involved in the study.

Data Availability Statement: The data presented in this study are available on valid request from the corresponding authors. The data are not publicly available due to privacy and ethical concerns.

Acknowledgments: This work was supported by Balneal and Rehabilitation Sanatorium Techirghiol.

Conflicts of Interest: The authors declare no conflict of interest.

References

1. Walston, J.D. Sarcopenia in Older Adults. *Curr. Opin. Rheumatol.* **2012**, *24*, 623–627. [CrossRef] [PubMed]
2. Yu, S.C.Y.; Khow, K.S.F.; Jadczak, A.D.; Visvanathan, R. Clinical Screening Tools for Sarcopenia and Its Management. *Curr. Gerontol. Geriatr. Res.* **2016**, *2016*, e5978523. [CrossRef] [PubMed]
3. Dhillon, R.J.S.; Hasni, S. Pathogenesis and Management of Sarcopenia. *Clin. Geriatr. Med.* **2017**, *33*, 17–26. [CrossRef] [PubMed]
4. Kara, M.; Kaymak, B.; Frontera, W.; Ata, A.M.; Ricci, V.; Ekiz, T.; Chang, K.-V.; Han, D.-S.; Michail, X.; Quittan, M.; et al. Diagnosing Sarcopenia: Functional Perspectives and a New Algorithm from the ISarcoPRM. *J. Rehabil. Med.* **2021**, *53*, jrm00209. [CrossRef] [PubMed]
5. Matsushita, T.; Nishioka, S.; Taguchi, S.; Yamanouchi, A.; Okazaki, Y.; Oishi, K.; Nakashima, R.; Fujii, T.; Tokunaga, Y.; Onizuka, S. Effect of Improvement in Sarcopenia on Functional and Discharge Outcomes in Stroke Rehabilitation Patients. *Nutrients* **2021**, *13*, 2192. [CrossRef] [PubMed]
6. Nishioka, S.; Matsushita, T.; Yamanouchi, A.; Okazaki, Y.; Oishi, K.; Nishioka, E.; Mori, N.; Tokunaga, Y.; Onizuka, S. Prevalence and Associated Factors of Coexistence of Malnutrition and Sarcopenia in Geriatric Rehabilitation. *Nutrients* **2021**, *13*, 3745. [CrossRef]
7. Remelli, F.; Vitali, A.; Zurlo, A.; Volpato, S. Vitamin D Deficiency and Sarcopenia in Older Persons. *Nutrients* **2019**, *11*, 2861. [CrossRef]
8. Yang, Y.J.; Kim, D.J. An Overview of the Molecular Mechanisms Contributing to Musculoskeletal Disorders in Chronic Liver Disease: Osteoporosis, Sarcopenia, and Osteoporotic Sarcopenia. *Int. J. Mol. Sci.* **2021**, *22*, 2604. [CrossRef]
9. Anker, S.D.; Morley, J.E.; von Haehling, S. Welcome to the ICD-10 Code for Sarcopenia. *J. Cachexia Sarcopenia Muscle* **2016**, *7*, 512–514. [CrossRef]
10. Stanciu, L.E.; Pascu, E.I.; Ionescu, E.V.; Circo, E.; Oprea, C.; Iliescu, M.G. Anti-ageing potential of techirghiol mud therapy through the modulation of pituitary- adrenal axis activity. *J. Environ. Prot. Ecol.* **2017**, *18*, 728–736.
11. Gupta, P.; Kumar, S. Sarcopenia and Endocrine Ageing: Are They Related? *Cureus* **2022**, *14*, e28787. [CrossRef] [PubMed]
12. Fan, J.; Kou, X.; Yang, Y.; Chen, N. MicroRNA-Regulated Proinflammatory Cytokines in Sarcopenia. *Mediat. Inflamm.* **2016**, *2016*, 1438686. [CrossRef] [PubMed]
13. Stangl, M.K.; Böcker, W.; Chubanov, V.; Ferrari, U.; Fischereder, M.; Gudermann, T.; Hesse, E.; Meinke, P.; Reincke, M.; Reisch, N.; et al. Sarcopenia—Endocrinological and Neurological Aspects. *Exp. Clin. Endocrinol. Diabetes* **2019**, *127*, 8–22. [CrossRef]
14. .Silva, J.; Martins, J.; Nicomédio, C.; Gonçalves, C.; Palito, C.; Gonçalves, R.; Fernandes, P.O.; Nunes, A.; Alves, M.J. A Novel Approach to Assess Balneotherapy Effects on Musculoskeletal Diseases—An Open Interventional Trial Combining Physiological Indicators, Biomarkers, and Patients' Health Perception. *Geriatrics* **2023**, *8*, 55. [CrossRef]
15. Ladang, A.; Beaudart, C.; Reginster, J.-Y.; Al-Daghri, N.; Bruyère, O.; Burlet, N.; Cesari, M.; Cherubini, A.; da Silva, M.C.; Cooper, C.; et al. Biochemical Markers of Musculoskeletal Health and Aging to Be Assessed in Clinical Trials of Drugs Aiming at the Treatment of Sarcopenia: Consensus Paper from an Expert Group Meeting Organized by the European Society for Clinical and Economic Aspects of Osteoporosis, Osteoarthritis and Musculoskeletal Diseases (ESCEO) and the Centre Académique de Recherche et d'Expérimentation En Santé (CARES SPRL), Under the Auspices of the World Health Organization Collaborating Center for the Epidemiology of Musculoskeletal Conditions and Aging. *Calcif. Tissue Int.* **2023**, *112*, 197–217. [CrossRef] [PubMed]

16. Delgado, D.A.; Lambert, B.S.; Boutris, N.; McCulloch, P.C.; Robbins, A.B.; Moreno, M.R.; Harris, J.D. Validation of Digital Visual Analog Scale Pain Scoring with a Traditional Paper-based Visual Analog Scale in Adults. *JAAOS Glob. Res. Rev.* **2018**, *2*, e088. [CrossRef] [PubMed]

17. Colleluori, G.; Villareal, D.T. Aging, Obesity, Sarcopenia and the Effect of Diet and Exercise Intervention. *Exp. Gerontol.* **2021**, *155*, 111561. [CrossRef] [PubMed]

18. Angulo, J.; El Assar, M.; Álvarez-Bustos, A.; Rodríguez-Mañas, L. Physical Activity and Exercise: Strategies to Manage Frailty. *Redox Biol.* **2020**, *35*, 101513. [CrossRef]

19. Smith, C.; Woessner, M.N.; Sim, M.; Levinger, I. Sarcopenia Definition: Does It Really Matter? Implications for Resistance Training. *Ageing Res. Rev.* **2022**, *78*, 101617. [CrossRef]

20. Moore, S.A.; Hrisos, N.; Errington, L.; Rochester, L.; Rodgers, H.; Witham, M.; Sayer, A.A. Exercise as a Treatment for Sarcopenia: An Umbrella Review of Systematic Review Evidence. *Physiotherapy* **2020**, *107*, 189–201. [CrossRef]

21. Raud, B.; Lanhers, C.; Crouzet, C.; Eschalier, B.; Bougeard, F.; Goldstein, A.; Pereira, B.; Coudeyre, E. Identification of Responders to Balneotherapy among Adults over 60 Years of Age with Chronic Low Back Pain: A Pilot Study with Trajectory Model Analysis. *Int. J. Environ. Res. Public. Health* **2022**, *19*, 14669. [CrossRef] [PubMed]

22. Popa, F.-L.; Iliescu, M.G.; Stanciu, M.; Georgeanu, V. Rehabilitation in a Case of Severe Osteoporosis with Prevalent Fractures in a Patient Known with Multiple Sclerosis and Prolonged Glucocorticoid Therapy. *Balneo PRM Res. J.* **2021**, *12*, 284–288. [CrossRef]

23. Popa, F.L.; Stanciu, M.; Bighea, A.; Berteanu, M.; Totoianu, I.G.; Rotaru, M. Decreased Serum Levels of Sex Steroids Associated with Osteoporosis in a Group of Romanian Male Patients. *Rev. Romana Med. Lab.* **2016**, *24*, 75–82. [CrossRef]

24. Popa, F.L.; Boicean, L.C.; Iliescu, M.G.; Stanciu, M. The Importance of Association between Sexsteroids Deficiency, Reduction of Bone Mineral Density and Falling Risk in Men with Implications in Medical Rehabilitation. *Balneo PRM Res. J.* **2021**, *12*, 318–322. [CrossRef]

25. Stanciu, M.; Boicean, L.C.; Popa, F.L. The Role of Combined Techniques of Scintigraphy and SPECT/CT in the Diagnosis of Primary Hyperparathyroidism: A Case Report. *Medicine* **2019**, *98*, e14154. [CrossRef]

26. Popa, F.L.; Stanciu, M.; Banciu, A.; Berteanu, M. Association between low bone mineral density, metabolic syndrome and sex steroids deficiency in men. *Acta Endocrinol.* **2016**, *12*, 418–422. [CrossRef]

 *medicina*

Article

Presence of Differences in the Radiofrequency Parameters Applied to Complex Pressure Ulcers: A Secondary Analysis

Miguel Ángel Barbas-Monjo [1,10], Eleuterio A. Sánchez-Romero [2,3,4,5,6,*], Jorge Hugo Villafañe [7], Lidia Martínez-Rolando [8], Jara Velasco García Cuevas [1] and Juan Nicolás Cuenca-Zaldivar [9,10,11]

1 Functional Recovery Unit, Guadarrama Hospital, 28440 Madrid, Spain
2 Department of Physiotherapy, Faculty of Sport Sciences, Universidad Europea de Madrid, 28670 Villaviciosa de Odón, Spain
3 Physiotherapy and Orofacial Pain Working Group, Sociedad Española de Disfunción Craneomandibular y Dolor Orofacial (SEDCYDO), 28009 Madrid, Spain
4 Musculoskeletal Pain and Motor Control Research Group, Faculty of Sport Sciences, Universidad Europea de Madrid, 28670 Villaviciosa de Odón, Spain
5 Department of Physiotherapy, Faculty of Health Sciences, Universidad Europea de Canarias, 38300 Santa Cruz de Tenerife, Spain
6 Musculoskeletal Pain and Motor Control Research Group, Faculty of Health Sciences, Universidad Europea de Canarias, 38300 Santa Cruz de Tenerife, Spain
7 IRCCS Fondazione Don Carlo Gnocchi, 20148 Milan, Italy
8 Rey Juan Carlos University Hospital of Móstoles, 28933 Madrid, Spain
9 Departamento de Enfermería y Fisioterapia, Grupo de Investigación en Fisioterapia y Dolor, Universidad de Alcalá, Facultad de Medicina y Ciencias de la Salud, 28801 Alcalá de Henares, Spain
10 Research Group in Nursing and Health Care, Puerta de Hierro Health Research Institute—Segovia de Arana (IDIPHISA), 28222 Majadahonda, Spain
11 Primary Health Center "El Abajón", Las Rozas de Madrid, 28231 Madrid, Spain
* Correspondence: eleuterio.sanchez@universidadeuropea.es; Tel.: +34-617-123-563; Fax: +34-633-115-328

Citation: Barbas-Monjo, M.Á.; Sánchez-Romero, E.A.; Villafañe, J.H.; Martínez-Rolando, L.; Velasco García Cuevas, J.; Cuenca-Zaldivar, J.N. Presence of Differences in the Radiofrequency Parameters Applied to Complex Pressure Ulcers: A Secondary Analysis. *Medicina* **2023**, *59*, 516. https://doi.org/10.3390/medicina59030516

Academic Editors: Milica Lazovic and Dejan Nikolić

Received: 9 January 2023
Revised: 21 February 2023
Accepted: 2 March 2023
Published: 7 March 2023

Abstract: *Background*: Pressure ulcers are a public health problem given the impact that they have on morbidity, mortality and the quality of life and participation of patients who suffer from them. Therefore, the main objective of this study was to evaluate the presence of differences in the radiofrequency parameters applied to complex pressure ulcers throughout the sessions and between the right and left leg. As a secondary objective, the subjective perceptions of the effects of the treatment by both the patients and the practitioner were analyzed. *Methods*: We performed a secondary analysis of data from a prospective study involving 36 patients from the Hospital de Guadarrama in Madrid, Spain, who presented ulcers in the lower limbs. Ten treatment sessions of radiofrequency were administered with a frequency of one session/week, collecting the data referring to the variables in each of the sessions. The main outcome variables were the radiofrequency parameters automatically adjusted in each session and that referred to the frequency (Hz), maximum and average power (W), absorbed energy by the ulcer (J/cm2) and temperature (°C) reached by the tissues. On the other hand, the subjective perception of the results was evaluated using the Global Response Assessment (GRA), a Likert-type scale that scores the treatment results from 1 (significantly worse) to 5 (significantly better). Likewise, the satisfaction of both the patients and the professional were evaluated using a 10-point numerical scale. *Results*: The ANOVA test showed significant differences ($p < 0.05$) throughout the sessions except in patient satisfaction. The ANOVA test showed significant differences ($p < 0.05$) between both legs and over time in all parameters except for frequency. The presence of significant differences ($p < 0.05$) was observed over time between legs compared to the initial values in the absorbed energy and in temperature, with higher final values in the absorbed energy in the left leg compared to the right (26.31 ± 3.75 W vs. 17.36 ± 5.66 W) and a moderate effect on both ($R^2 = 0.471$ and 0.492, respectively). The near absence of changes in the satisfaction of both the patients and the professional was observed, while the score in the GRA decreased continuously throughout the sessions. *Conclusions*: Radiofrequency parameters are indicative of an improved clinical response to ulcers. In addition, higher radiofrequency exposure increases healing capacity. However, the subjective perception of treatment outcomes worsened, which may be related to the chronic nature of the ulcers, leading to patients' expectations not being met.

Keywords: pressure ulcers; radiofrequency; expectations

1. Introduction

Pressure ulcers are a public health problem given the impact that they have on morbidity, mortality and the quality of life and participation of patients who suffer with them [1]. The prevalence in Spain is around 7.9% in adults, whereby 65.6% of which are of nosocomial origin; in the United States, they are suffered by some 2.5 million individuals, with percentages ranging between 5% and 15% [1,2]. The classification of these injuries is based on tissue damage from level I with superficial red areas to level IV with significant skin damage that may involve bone, tendon or joint capsule [3]. These hospital-acquired pressure injuries (HAPIs) can lead to chronic wounds, contractures, osteomyelitis, loss of limbs and sepsis, and cause about 60,000 deaths per year [2].

The annual cost of this type of injury is between USD 3.3 billion and USD 11 billion per year and can be as high as USD 26.8 billion in the most advanced stages of HAPI, making the cost per patient about USD 10,708, and each hospital episode costs between USD 500 and more than USD 70,000 in the United States [2].

In diabetic patients, the prevalence of distal ulcers in the lower limbs is between 19% and 34%, being one of the main complications of this pathology. These lesions account for one third of the diabetic patient's expenses, reaching a total annual expenditure of USD 176 billion; however, despite this high cost, 20% of patients continue to have ulcers after one year of treatment [4].

Given the impact on people's lives and the cost that pressure ulcers and their complications represent for the healthcare system, prevention should be the most efficient method for dealing with them [1], considering taking measures to address risk factors such as limited movement, the elderly and prolonged embedding [5]. However, the high prevalence makes it necessary to develop and implement treatment methods that limit the process by accelerating recovery and avoiding the appearance of the complications.

In healthy subjects without affectation of the sensory, motor and mental areas, the maintenance of static positions leads to posture modification; however, this does not happen in those patients with alteration of any of the mentioned spheres, causing pressures higher than the filling pressure of the arterial capillaries and higher than the outflow pressure of the venous capillaries to produce tissue hypoxia, resulting in ischemia, tissue damage and subsequent necrosis [6,7].

These lesions can take decades to heal or even fail to heal, which can have an impact on people's lives, leading to the appearance of secondary diseases such as depression or family distress [8]; however, there are no studies on electrotherapy that analyze factors such as quality of life, depression or the perceived effectiveness of the treatment [3]. Although, it has been evidenced that patients understand that the improvement in the lesion is due to a collaborative approach where they feel more knowledgeable and empowered with tools to improve ulcer care [9].

The use of electrotherapy to address these lesions has been widely studied, especially using electrical stimulation with involvement in the four phases of healing (inflammatory, proliferative, epithelialization and remodeling phase); although the mechanism of action is not well understood, the evidence suggests that this practice increases the flow of blood and thus that of cells, promotes oxygenation, reduces edema and influences dermal growth factors and their receptors [3,10].

Although not as widely studied, radiofrequency has also been used as part of the treatment of HAPI; it began to be used in 1950 and continues to be used today with the aim of improving the remodeling of the injured tissue without causing damage to the surrounding healthy tissue [11].

According to the literature related to the use of radiofrequency in the treatment of chronic pressure ulcers, several studies reflect the use of pulsed radiofrequency as part of the multimodal treatment of these difficult lesions [10,12,13]. These show that this type of treatment, being non-invasive, relatively inexpensive, easy and safe to use and with good patient acceptance is an interesting tool to use as an adjuvant treatment [10]. In relation to multimodal therapy, radiofrequency is combined both with conventional therapy (care appropriate to the lesion) and with negative pressure devices and/or dermal replacement [10,12,13]. The evidence shows the usefulness of pulsed radiofrequency as part of the intervention, achieving progressive healing and thus avoiding amputation of the affected limb [12,13].

Other authors mention a variant of pulsed radiofrequency called pulse dose radiofrequency, where the constant is determined by the voltage and not by the time, as in pulsed radiofrequency, which ensures that the tissue temperature is maintained at 42 °C and allows standardizing a treatment method in relation to the dose and not the exposure time, where preliminary results in small samples show a better effectiveness in reducing pain and maintaining the results achieved [14].

Tecartherapy, included in this group, consists of endogenous thermotherapy using electric current through monopolar capacitive and resistive radiofrequency with the aim of producing heat in the most superficial tissues (capacitive electrode) and in the deepest ones (resistive electrode) [11]. The heat produced in the tissues generates a tissue damage that stimulates fibroblasts and growth factors favoring the production and remodeling of collagen and elastin and the deposition of hyaluronic acid de novo, obtaining as a final result a thickening of the subcutaneous tissue layer that avoids necrosis, fibrosis and damage to vascular and adnexal structures [15].

The main objective of this study was to evaluate the presence of differences between the radiofrequency parameters applied to complex pressure ulcers throughout the sessions and between the right and left leg. As a secondary objective, the subjective perception of the effects of the treatment by both the patients and the practitioner was analyzed.

2. Materials and Methods

2.1. Study Design

We performed a secondary analysis of data from a prospective study. The previous study aimed to evaluate the effect that the application of radiofrequency at low intensity (frequency) and with non-thermal effects has on the different components of the mechanism of the healing process of hard-to-heal lesions. The methods and description of the study have been previously described [16]. The current study was approved by the Clinical Research Ethics Committee of the Puerta de Hierro Hospital, Madrid, Spain (approval number: 02.18, 6 February 2018). All patients provided informed consent prior to their enrollment. The most relevant parts of the design are summarized below.

2.2. Study Population

The study included 36 patients from the Hospital de Guadarrama in Madrid, Spain, who presented ulcers in the lower limbs considering the following aspects of their case histories: age, height and weight, and 13 men and 10 women had diabetes mellitus. These patients came directly from the hospital admission service for ulcer consultation, where they were evaluated by a geriatrician with 20 years of experience (J.V.G.C.). The research team took into account the following inclusion criteria: male or female over or equal to 18 years old or under or equal to 90 years old with a diagnosis of a long-lasting complex wound, that admission was the first one for treatment at Hospital de Guadarrama and that the patient understood and voluntarily signed the corresponding informed consent sheet and information sheet prior to the performance of any evaluation or procedure related to the study. Patients with the following comorbidities were excluded from the study: cardiac pacemaker wearers, presence of local metallic implants, lesion infection, patients with cognitive impairment and patients with malnutrition or risk of malnutrition.

2.3. Outcome Measures

The main outcome variables were the radiofrequency parameters automatically adjusted in each session and that referred to the frequency (Hz), maximum and average power (W), absorbed energy by the ulcer (J/cm^2) and temperature (°C) reached by the tissues.

On the other hand, the subjective perception of the results was evaluated using the Global Response Assessment (GRA), a Likert-type scale that scores the treatment results from 1 (significantly worse) to 5 (significantly better) [11,17]. Likewise, the satisfaction of both the patients and the professional were evaluated using a 10-point numerical scale.

Ten treatment sessions were administered with a frequency of one session/week, with all outcome measures measured at each treatment session.

2.4. Intervention

Treatment was administered by M.A.B.M., a nurse specialist with 30 years of experience in ulcer treatment, using the CAPENERGY Vascular C200 (CE120) tecartherapy device with a C-Boot foot probe in the case of lesions on the sole of the foot, or with capacitive plates in the rest of the body areas.

A total of 10 radiofrequency sessions were applied in the 36 patients with a periodicity of once a week, with a power of 60% and a frequency of 1.2 MHz for 30 min, placing the treatment head on the lesion, and an athermal dose of up to 37 °C was administered (Figure 1).

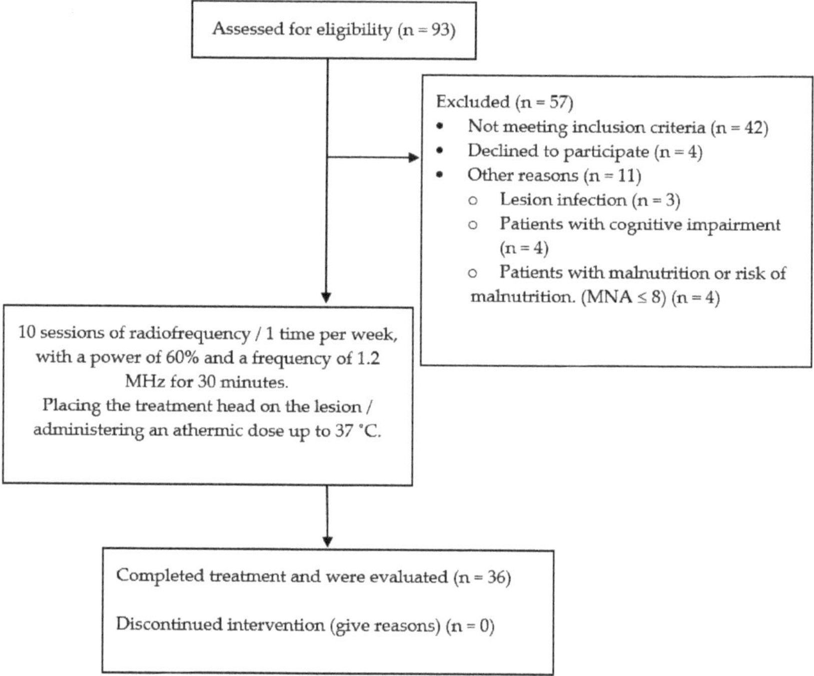

Figure 1. Flow chart of the study.

2.5. Statistical Analysis

Statistical analysis was performed using the program R Ver. 3.5.1 (R Foundation for Statistical Computing, Institute for Statistics and Mathematics, Welthandelsplatz 1, 1020 Vienna, Austria). The significance level was set at $p < 0.05$. The distribution of quantitative variables was tested using the Shapiro–Wilk test which evidenced the absence of normality. Qualitative variables were described in absolute values and frequencies and quantitative variables were described using mean and standard deviation.

Given that this is a single-group observational study, the sample size was not calculated a priori, but included patients who attended the ulcer consultation at Hospital Guadarrama (after signing the informed consent and meeting the eligibility criteria) during the period from September 2018 to June 2019. The final power of the study was calculated using the program R Ver. 3.5.1 (R Foundation for Statistical Computing, Institute for Statistics and Mathematics, Welthandelsplatz 1, 1020 Vienna, Austria), applying the Wilcoxon signed-rank test with Bonferroni correction with the PUSH scale scores between the first and last session.

Changes over the sessions in subjective perception and radiofrequency parameters were analyzed and, in the case of radiofrequency parameters, the differences between the right and left legs over the sessions were also analyzed. In both cases, a repeated measures linear mixed model and restricted maximum likelihood (REML) and unstructured correlation (default) structure were used. The subjects were modeled according to random effect and time in the first case, or the group (leg): time interaction in the second as fixed effects, adjusting in the latter case the results with the baseline values. Due to the small sample size, the Kenward–Roger degrees of freedom correction was applied and confidence intervals were calculated via bootstrap. The Nakagawa and Schielzeth R^2 was calculated for each model as a goodness-of-fit measure. Post hoc matched pair comparisons were applied using the Bonferroni correction.

3. Power Analysis

Accepting a risk α of 0.05, the final power of the study was estimated at 100% with a final mean PUSH score of 10.695 at the first session and 4.695 at the ending treatment session.

4. Results

The sample was composed of 36 subjects of 63.31 $\pm$ 9.99 years, with a majority of women (58.3%) and with risk factors such as diabetes mellitus (66.7%), dyslipidemia (72.2%) or high blood pressure (75.0%) (Table 1).

Table 1. Clinical and demographic characteristics of the participants.

		36
n		36
Age		63.31 $\pm$ 9.9
Weight		72.25 $\pm$ 8.8
Height		162.33 $\pm$ 3.8
Gender, n(%)	Female	21 (58)
	Male	15 (41)
Smoking, n(%)	No	36 (100)
Embolism, n(%)	No	36 (100)
Diabetes mellitus, n(%)	No	12 (33)
	Yes	24 (66)
Dyslipidemia, n(%)	No	10 (28)
	Yes	26 (72)
Arterial hypertension, n(%)	No	9 (25)
	Yes	27 (75)
Kidney failure, n(%)	No	34 (94)
	Yes	2 (6)
Parkinson's, n(%)	No	36 (100)
Alzheimer's, n(%)	No	36 (100)
Dementia, n(%)	No	36 (100)
Respiratory pathology, n(%)	No	36 (100)
Cardiac pathology, n(%)	No	36 (100)
Musculoskeletal pathology, n(%)	No	36 (100)
Other pathologies, n(%)	No	36 (100)
Nutritional supplements, n(%)	No	36 (100)
Muscle relaxants, n(%)	No	36 (100)
Anxiolytics, n(%)	No	35 (97)
	Yes	1 (3)

Table 1. *Cont.*

Antiepileptics, n(%)	No	34 (94)
	Yes	2 (6)
Analgesics, n(%)	No	30 (83.3)
	Yes	6 (17)
Opioids, n(%)	No	36 (100)
Statins, n(%)	No	36 (100)
Antiaggregants, n(%)	No	36 (100)
Fever, n(%)	No	36 (100.0)
Pain, n(%)	No	30 (83.3)
	Yes	6 (16.7)
Traumatism, n(%)	No	36 (100.0)

Data expressed as mean ± standard deviation or with absolute and relative values (%).

The recruited patients presented pressure ulcers with an average surface of 25 cm^2 grade III-IV with an evolution time of approximately 2 months with venous vascular lesions in women and arterial lesions in men.

4.1. Comparisons across Sessions

The ANOVA test showed significant differences ($p < 0.05$) throughout the sessions, except in patient satisfaction. It was verified that the presence of systematic differences between practically all of the measurement moments compared with the initial values and specifically between the final values and the initial ones, except in the satisfaction of the patients and the professional and in the frequency in the left leg ($p < 0.05$) (Table 2 and Supplementary Material Table S1).

These differences were translated into a final increase in the values of the radio frequency parameters in both legs and a decrease in the Global Response Assessment score with very high effects on the latter (R2 = 0.874) and being greater than 0.5 in the parameters of the left leg (Supplementary Material Table S2).

The pairwise comparisons showed systematic differences between practically all of the measurement moments, with significant differences being observed in all of the variables between the first and the last session ($p < 0.05$), except in the satisfaction of the patients and the professional (Supplementary Material Table S3).

4.2. Comparison of Radiofrequency Parameters between Both Legs

The ANOVA test showed significant differences ($p < 0.05$) between both legs and over time in all parameters except for frequency. The presence of significant differences ($p < 0.05$) was observed over time between legs compared to the initial values in the absorbed energy and in temperature, with higher final values in the absorbed energy in the left leg compared to the right (26.31 ± 3.75 W vs. 17.36 ± 5.66 W) and a moderate effect on both (R2 = 0.471 and 0.492, respectively) (Table 3 and Supplementary Material Table S4).

Table 2. Subjective perception and radiofrequency parameters: model results.

	Global Response Assessment	Patients' Satisfaction	Professional Satisfaction	Right Frequency (Hz)	Right Maximum Power (Watts)	Right Average Power (Watts)	Right Absorbed Energy (Volt-Ampere)	Right Temperature (Celsius)	Left Frequency (Hz)	Left Maximum Power (Watts)	Left Average Power (Watts)	Left Absorbed Energy (Volt-Ampere)	Left Temperature (Celsius)
Nakagawa and Schielzeth R^2	0.874	0.397	0.437	0.081	0.435	0.276	0.401	0.344	0.687	0.676	0.564	0.708	0.729
T10-T1 difference (95%CI)	−2.861 (−2.98, −2.742)	−0.071 (−0.333, 0.19)	−0.024 (−0.224, 0.177)	14.917 (12.698, 17.136)	13.612 (10.063, 17.162)	8.596 (6.009, 11.183)	6.03 (5.142, 6.918)	−33.15 (−34.464, −31.837)	−0.03 (−0.076, 0.015)	21.956 (15.106, 28.805)	21.166 (17.509, 24.822)	25.333 (24.726, 25.939)	1.833 (0.792, 2.873)
Omnibus ANOVA: time	$F(9, 315) = 247.659, p = <0.001$	$F(9, 369) = 1.261, p = 0.257$	$F(9, 369) = 6.346, p = <0.001$	$F(9, 369) = 4.126, p = <0.001$	$F(9, 369) = 11.563, p = <0.001$	$F(9, 369) = 6.852, p = <0.001$	$F(9, 369) = 6.919, p = <0.001$	$F(9, 369) = 16.484, p = <0.001$	$F(9, 81) = 15.793, p = <0.001$	$F(9, 81) = 20.876, p = <0.001$	$F(9, 81) = 13.503, p = <0.001$	$F(9, 81) = 26.622, p = <0.001$	$F(9, 81) = 25.628, p = <0.001$

F: F statistic (degrees of freedom); 95%CI: 95% confidence interval. Significant if $p < 0.05$ (shown in red).

Table 3. Radiofrequency parameters between both legs: model results.

	Frequency (Hz)	Maximum Power (Watts)	Average Power (Watts)	Absorbed Energy (Volt-Ampere)	Temperature (Celsius)
Nakagawa and Schielzeth R^2	0.058	0.522	0.301	0.471	0.492
Omnibus ANOVA: group:time	$F_{(8, 412.104)} = 0.368$, $p = 0.937$	$F_{(8, 409.049)} = 8.882$, $p = <0.001$	$F_{(8, 411.292)} = 5.702$, $p = <0.001$	$F_{(8, 409.197)} = 8.484$, $p = <0.001$	$F_{(8, 409.175)} = 12.53$, $p = <0.001$
T10 between groups' adjusted difference (95%CI)	0.2 (−1.61, 2.01)	−0.3 (−1.421, 0.821)	−0.02 (−0.065, 0.025)	4.3 (2.304, 6.296)	<0.001 (< 0.001, <0.001)

F: F statistic (degrees of freedom); 95%CI: 95% confidence interval; 95%CI: 95% confidence interval. Significant if p < 0.05 (shown in red).

Pairwise comparisons showed differences in absorbed energy between both legs at the end, with no initial significant differences ($p = 0.049$), while significant differences in temperature occurred in the first treatment sessions (Supplementary Material Table S5).

All radiofrequency parameters were found to increase progressively throughout the sessions and more markedly in the left leg, especially with regard to the absorbed energy where the confidence intervals barely overlapped (Figure 2).

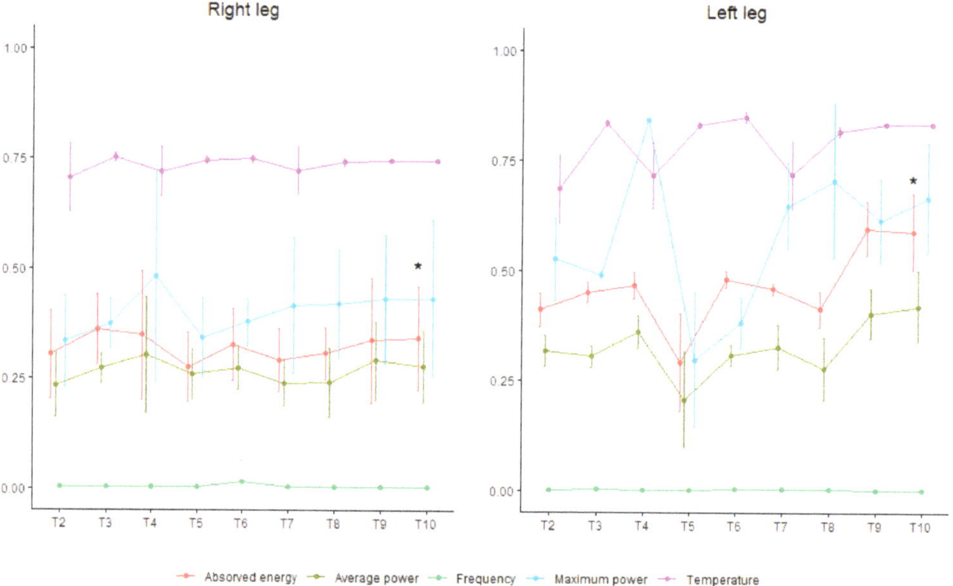

Figure 2. Radiofrequency parameters (standardized adjusted mean and standard deviation error bars). * Radiofrequency parameter significant differences between both legs.

4.3. Satisfaction of the Patients and the Professional

A near absence of changes in the satisfaction of both the patients and the professional was observed, while the score in the Global Response Assessment decreased continuously throughout the sessions (Figure 3).

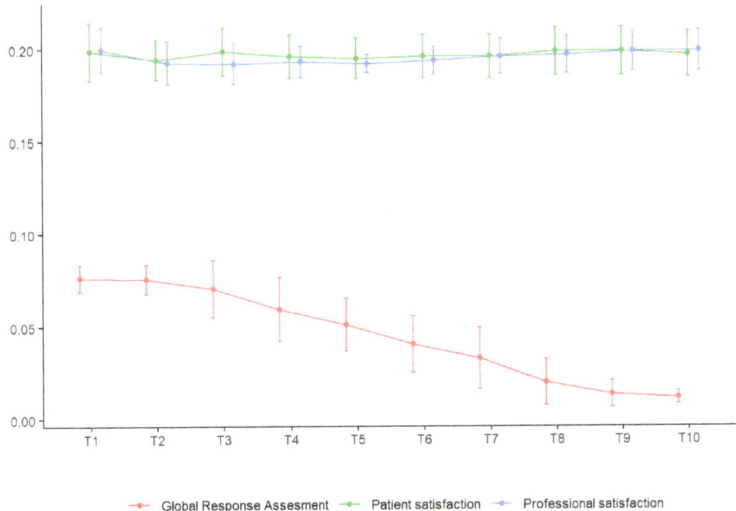

Figure 3. Subjective perception (standardized mean and standard deviation error bars).

5. Discussion

The results of the present study show that an increase in the radiofrequency parameters, including temperature and especially the absorbed energy, especially in the left leg, are indicative of an improvement in the clinical response of the ulcers. Furthermore, a greater exposure to radiofrequency increases the healing power [18–20]. It is important to detail that no relevant influences were found with respect to the differences in the results influenced by clinical variables and by sex or any other sociodemographic variables. These results also agree with the previous study [16] in which there was an average increase in temperature by thermography of 1.4 °C, as well as a healing rate percentage of 60% with a progressive reduction in size and exudate of the ulcer measured using the Pressure Ulcer Scale for Healing (PUSH). However, it has not been possible to establish the reason as to why the clinical response of the ulcers in the left leg was better than in the right leg and there are no other studies that have studied the differences between a lower limb and the contralateral limb using tecartherapy. In addition, a review in other fields of electrotherapy such as electrostimulation concluded that there is no consensus on parameters such as frequency, duration and location of treatment [3]. Future studies are therefore needed to study the appropriate dose and whether the differences between one limb and another may be due to tissue factors.

Likewise, for the subjective perception of the results by the patient (GRA), the scores were always higher in patients with dyslipidemia or arterial hypertension compared to those without. In the case of professional satisfaction, patients with renal failure reported higher scores than those without this pathology. We cannot establish hypotheses regarding these results for the moment, on the one hand due to the low sample size of the present study, and on the other hand due to the lack of studies on which to discuss these results. This is because, to the knowledge of the authors of the present study, this is the first time that such comparisons have been made using radiofrequency. This could certainly be a line of research, or at least parameters to be analyzed in future studies.

To our knowledge, this is the first study that uses this type of therapy to address pressure ulcers and assesses parameters of subjective perception of outcomes as well as patient and professional satisfaction. A Cochrane review about electrical stimulation for the treating of pressure ulcers exposes the absence of studies evaluating parameters such as quality of life, depression or perception of treatment effectiveness, despite these being relevant outcomes for patients [3]. In this study, the subjective perception of the results of the treatment measured with the GRA scale worsened, which may be related to the chronic nature of the ulcers and the fact that they did not disappear completely and perhaps did not respond to the expectations of the patients. In this regard, Wood et al. [9], in a study on the collaborative management of pressure ulcers, concluded that patients felt that pressure ulcer care improved by using a collaborative, multidisciplinary approach because they felt more knowledgeable, empowered and more able to improve their pressure ulcer care. This issue was approached from a qualitative perspective by García Sánchez et al. [21], concluding that proximity, trust and effective and bidirectional communication between patients and health professionals are fundamental. Due to a scarcity of studies assessing factors as relevant to the patient as those described above and considering the importance of the patient's perception in the treatment, future studies concerning these aspects are necessary.

In this sense, Ballestra et al. [22] showed that certain patient expectations, such as the expectation of a tailored treatment with frequent follow-ups, the hope of obtaining the best possible results, realism or resignation regarding the alleviation of the health problem, good dialogue and communication, the need to be seen and confirmed as an individual and the desire to receive an explanation of their disease, could be related to better recovery results.

However, it appears that the induction of different types of expectations (positive or negative) through verbal suggestion does not influence the perception of acute pain perceived during the performance of a technique that may be painful [23].

Limitations

An important limitation of this study is the small sample size. Additionally important is the absence of a control group or a placebo group to compare with the evolution of the process or with other interventions. The authors also recognize as a limitation the fact that they did not measure the psychological and behavioral factors of the sample analyzed in the study, which is observed as a determinant in different chronic diseases [24,25]. Although there is an inherent bias in the type of research model, this has been minimized by using relevant and reliable human (accredited clinical experience of the researchers) and instrumental resources.

6. Conclusions

It was observed that the radiofrequency parameters increased progressively throughout the sessions and more markedly in the left leg, but for the difference between legs it was not possible to establish the reason for this clinical response. This increase was indicative of an improvement in the clinical response to ulcers.

In addition, higher radiofrequency exposure increases healing capacity.

However, the subjective perception of treatment outcomes worsened, which may be related to the chronic nature of the ulcers, leading to patients' expectations not being met.

7. Key Points

In summary, considering the great clinical potential of radiofrequency, we can expect an increase in new techniques for tissue regeneration and wound healing in the near future.

The increased power to accelerate the wound healing process can be explained by the anti-inflammatory effect caused by the changes that occur in the perilesional skin, and the improvement in microcirculation contributes to the increase in the reactivity of the different layers of the skin.

The influence of the magnetic field on the microcirculatory system can be used to explain the often-cited fact that magnetic fields have anti-oedematous, analgesic and anti-inflammatory effects, which is one of the reasons for their wide application in the field of injury treatment.

The subjective perception of treatment success may be influenced by the chronic nature of the ulcers, which leads to patients' expectations not being met.

Supplementary Materials: The following supporting information can be downloaded at https://www.mdpi.com/article/10.3390/medicina59030516/s1: Table S1: Subjective perception and radiofrequency parameters model coefficient terms results; Table S2: Subjective perception and radiofrequency parameters outcomes; Table S3: Subjective perception and radiofrequency parameters overall pairwise model coefficient terms results; Table S4: Radiofrequency parameters between both legs model coefficient terms results; Table S5: Radiofrequency parameters between both legs model pairwise results.

Author Contributions: Conceptualization, M.Á.B.-M., J.V.G.C., J.N.C.-Z. and E.A.S.-R.; methodology, J.N.C.-Z. and E.A.S.-R.; software, J.N.C.-Z.; validation, all authors.; formal analysis, J.N.C.-Z. and E.A.S.-R.; investigation, all authors.; resources, J.N.C.-Z.; data curation, J.N.C.-Z.; writing—original draft preparation, J.N.C.-Z., L.M.-R. and E.A.S.-R.; writing—review and editing, J.N.C.-Z., L.M.-R., J.H.V. and E.A.S.-R.; visualization, E.A.S.-R. and J.N.C.-Z.; supervision, all authors.; project administration, J.N.C.-Z.; funding acquisition, J.N.C.-Z. All authors have read and agreed to the published version of the manuscript.

Funding: The APC was funded by Fundación de Investigation Biomedica del Hospital Puerta de Hierro de Majadahonda, C/Joaquín Rodrigo 2, Edif Laboratorios, Planta 0, 28222, Madrid.

Institutional Review Board Statement: The study was conducted in accordance with the Declaration of Helsinki and was approved by the Clinical Research Ethics Committee of the Puerta de Hierro Hospital, Madrid, Spain (approval number: 02.18).

Informed Consent Statement: Informed consent was obtained from all subjects involved in the study. At all times, the confidentiality of the information was preserved, making responsible use of the data, as established by current Spanish regulations and in accordance with the Declaration of Helsinki.

Data Availability Statement: The data presented in this study are available upon request from the corresponding authors. The data are not publicly available due to ethical restrictions.

Acknowledgments: This study was supported by the Italian Ministry of Health, Ricerca Corrente, 2023.

Conflicts of Interest: The authors declare no conflict of interest. The funders had no role in the design of the study, in the collection, analyses or interpretation of data, in the writing of the manuscript or in the decision to publish the results.

References

1. Quiñoz Gallardo, M.D.; Gonzalo Jiménez, E.; Barrientos Trigo, S.; Porcel Gálvez, A.M. Implantación del programa "Centros Comprometidos con la Excelencia en Cuidados" en el Hospital Universitario Virgen de las Nieves. *Rev. Española De Salud Pública* **2021**, *95*, e1–e16.
2. Padula, W.V.; Delarmente, B.A. The national cost of hospital-acquired pressure injuries in the United States. *Int. Wound J.* **2019**, *16*, 634–640. [CrossRef]
3. Arora, M.; Harvey, L.A.; Glinsky, J.V.; Nier, L.; Lavrencic, L.; Kifley, A.; Cameron, I.D. Electrical stimulation for treating pressure ulcers. *Cochrane Database Syst. Rev.* **2020**, *1*, CD012196. [CrossRef] [PubMed]
4. Everett, E.; Mathioudakis, N. Update on management of diabetic foot ulcers: Diabetic foot ulcers. *Ann. N. Y. Acad. Sci.* **2018**, *1411*, 153–165. [CrossRef] [PubMed]
5. Mao, X.; Zhu, L. Effects of care bundles for patients with pressure ulcers and the impact on self-care efficacy. *Am. J. Transl. Res.* **2021**, *13*, 1799–1807.
6. Mervis, J.S.; Phillips, T.J. Pressure ulcers: Pathophysiology, epidemiology, risk factors, and presentation. *J. Am. Acad. Dermatol.* **2019**, *81*, 881–890. [CrossRef]
7. Micah, A.E.; Cogswell, I.E.; Cunningham, B.; Ezoe, S.; Harle, A.C.; Maddison, E.R.; McCracken, D.; Nomura, S.; Simpson, K.E.; Stutzman, H.N.; et al. Tracking development assistance for health and for COVID-19: A review of development assistance, government, out-of-pocket, and other private spending on health for 204 countries and territories, 1990–2050. *Lancet* **2021**, *398*, 1317–1343. [CrossRef]
8. Bowers, S.; Franco, E. Chronic Wounds: Evaluation and Management. *Am. Fam. Physician* **2020**, *101*, 8.
9. Wood, J.; Brown, B.; Bartley, A.; Cavaco, A.M.B.C.; Roberts, A.P.; Santon, K.; Cook, S. Reducing pressure ulcers across multiple care settings using a collaborative approach. *BMJ Open Qual.* **2019**, *8*, e000409. [CrossRef]
10. Conner-Kerr, T.; Isenberg, R.A. Retrospective analysis of pulsed radiofrequency energy therapy use in the treatment of chronic pressure ulcers. *Adv. Ski. Wound Care* **2012**, *25*, 253–260. [CrossRef]

11. Ekelem, C.; Thomas, L.; Van Hal, M.; Valdebran, M.; Lotfizadeh, A.; Mlynek, K.; Mesinkovska, N.A. Radiofrequency Therapy and Noncosmetic Cutaneous Conditions. *Dermatol. Surg.* **2019**, *45*, 908–930. [CrossRef] [PubMed]

12. Frykberg, R.; Martin, E.; Tallis, A.; Tierney, E. A case history of multimodal therapy in healing a complicated diabetic foot wound: Negative pressure, dermal replacement and pulsed radio frequency energy therapies. *Int. Wound J.* **2011**, *8*, 132–139. [CrossRef] [PubMed]

13. Larsen, J.A.; Overstreet, J. Pulsed radio frequency energy in the treatment of complex diabetic foot wounds: Two cases. *J. Wound Ostomy Cont. Nurs.* **2008**, *35*, 523–527. [CrossRef] [PubMed]

14. Quarto, G.; Solimeno, G.; Furino, E.; Sivero, L.; Bucci, L.; Massa, S.; Benassai, G.; Apperti, M. "Difficult-to treat" ulcers management: Use of pulse dose radiofrequency. *Ann. Ital. Di Chir.* **2013**, *84*, 225–228.

15. Clijsen, R.; Leoni, D.; Schneebeli, A.; Cescon, C.; Soldini, E.; Li, L.; Barbero, M. Does the Application of Tecar Therapy Affect Temperature and Perfusion of Skin and Muscle Microcirculation? A Pilot Feasibility Study on Healthy Subjects. *J. Altern. Complement. Med.* **2020**, *26*, 147–153. [CrossRef]

16. Barbas Monjo, M.; Velasco García Cuevas, J.; Rodríguez Lastra, J.; Cuenca Zaldívar, J.N. Radiofrecuencia en la cicatrización de heridas crónicas. Una revisión en hospital de media estancia. *Gerokomos* **2021**, *32*, 63–67. [CrossRef]

17. Fernández-Carnero, J.; Beltrán-Alacreu, H.; Arribas-Romano, A.; Cerezo-Téllez, E.; Cuenca-Zaldivar, J.N.; Sánchez-Romero, E.A.; Lerma Lara, S.; Villafañe, J.H. Prediction of Patient Satisfaction after Treatment of Chronic Neck Pain with Mulligan's Mobilization. *Life* **2023**, *13*, 48. [CrossRef]

18. Peters, K.M.; Killinger, K.A.; Ibrahim, I.A.; Villalba, P.S. The relationship between subjective and objective assessments of sacral neuromodulation effectiveness in patients with urgency-frequency. *Neurourol. Urodyn.* **2008**, *27*, 775–778. [CrossRef]

19. Liburdy, R.P.; Callahan, D.E.; Harland, J.; Dunham, E.; Sloma, T.R.; Yaswen, P. Experimental evidence for 60 Hz magnetic fields operating through the signal transduction cascade. Effects on calcium influx and c-MYC mRNA induction. *FEBS Lett.* **1993**, *334*, 301–308. [CrossRef]

20. Yokota, Y.; Tashiro, Y.; Suzuki, Y.; Tasaka, S.; Matsushita, T.; Matsubara, K.; Kawagoe, M.; Sonoda, T.; Nakayama, Y.; Hasegawa, S.; et al. Effect of Capacitive and Resistive Electric Transfer on Tissue Temperature, Muscle Flexibility, and Blood Circulation. *J. Nov. Physiother.* **2017**, *7*, 325. [CrossRef]

21. García-Sánchez, F.J.; Martínez-Vizcaíno, V.; Rodríguez-Martín, B. Patients' and Caregivers' Conceptualisations of Pressure Ulcers and the Process of Decision-Making in the Context of Home Care. *Int. J. Environ. Res. Public Health.* **2019**, *16*, 2719. [CrossRef]

22. Ballestra, E.; Battaglino, A.; Cotella, D.; Rossettini, G.; Sanchez-Romero, E.A.; Villafane, J.H. ¿Influyen las expectativas de los pacientes en el tratamiento conservador de la lumbalgia crónica? Una revisión narrativa (Do patients' expectations influence conservative treatment in Chronic Low Back Pain? A Narrative Review). *Retos* **2022**, *46*, 395–403. [CrossRef]

23. Sánchez Romero, E.A.; Lim, T.; Villafañe, J.H.; Boutin, G.; Riquelme Aguado, V.; Martin Pintado-Zugasti, A.; Alonso Pérez, J.L.; Fernández Carnero, J. The Influence of Verbal Suggestion on Post-Needling Soreness and Pain Processing after Dry Needling Treatment: An Experimental Study. *Int. J. Environ. Res. Public Health* **2021**, *18*, 4206. [CrossRef]

24. Viñals Narváez, A.C.; Sánchez-Sánchez, T.; García-González, M.; Ardizone García, I.; Cid-Verdejo, R.; Sánchez Romero, E.A.; Jiménez-Ortega, L. Psychological and Behavioral Factors Involved in Temporomandibular Myalgia and Migraine: Common but Differentiated Profiles. *Int. J. Environ. Res. Public Health* **2023**, *20*, 1545. [CrossRef] [PubMed]

25. Osses-Anguita, Á.E.; Sánchez-Sánchez, T.; Soto-Goñi, X.A.; García-González, M.; Alén Fariñas, F.; Cid-Verdejo, R.; Sánchez Romero, E.A.; Jiménez-Ortega, L. Awake and Sleep Bruxism Prevalence and Their Associated Psychological Factors in First-Year University Students: A Pre-Mid-Post COVID-19 Pandemic Comparison. *Int. J. Environ. Res. Public Health* **2023**, *20*, 2452. [CrossRef] [PubMed]

Article

Morphometric Parameters and MRI Morphological Changes of the Knee and Patella in Physically Active Adolescents

Goran Djuricic [1,2,*], Filip Milanovic [3], Sinisa Ducic [1,3], Vladimir Radlović [1,3], Mikan Lazovic [3], Ivan Soldatovic [1] and Dejan Nikolic [1,4]

1 Faculty of Medicine, University of Belgrade, 11000 Belgrade, Serbia
2 Radiology Department, University Children's Hospital, 11000 Belgrade, Serbia
3 Orthopedic Surgery and Traumatology Department, University Children's Hospital, 11000 Belgrade, Serbia
4 Physical Medicine and Rehabilitation Department, University Children's Hospital, 11000 Belgrade, Serbia
* Correspondence: gorandjuricic@gmail.com

Abstract: *Background and Objectives*: The immature skeleton in a pediatric population exposed to frequent physical activity might be extremely prone to injuries, with possible consequences later in adulthood. The main aim of this study is to present specific morphometric parameters and magnetic resonance imaging (MRI) morphological changes of the knee and patella in a physically active pediatric population. Additionally, we wanted to investigate the morphological risk factors for patellar instability. *Materials and Methods:* The study included the MRI findings of 193 physically active pediatric patients with knee pain. The participants underwent sports activities for 5 to 8 h per week. Two divisions were performed: by age and by patellar type. We evaluated three age groups: group 1 (age 11–14), group 2 (age 15–17), and group 3 (age 18–21 years). In addition, participants were divided by the patellar type (according to Wiberg) into three groups. The following morphometric parameters were calculated: lateral trochlear inclination (LTI), the tibial tubercle–trochlear groove distance (TT-TG), trochlear facet asymmetry (TFA), Insall–Salvati index, modified Insall–Salvati index, Caton–Deschamps index, articular overlap, morphology ratio and contact surface ratio. *Results:* We found a statistically significant association between patellar type groups in LTI ($p < 0.001$), TFA ($p < 0.001$), Insal–Salvati ($p = 0.001$) index, and Caton–Deschamps index ($p = 0.018$). According to age groups, we found statistical significance in the Caton–Deschamps index ($p = 0.039$). The most frequent knee injury parameter, according to Wiberg, in physically active pediatric patients was patella type 2 in boys and type 3 in girls. *Conclusions:* The MRI morphometric parameters observed in our study might be factors of prediction of knee injury in physically active children. In addition, it might be very useful in sports programs to improve the biomechanics of the knee in order to reduce the injury rate in sports-active children.

Keywords: knee injury; magnetic resonance imaging; children; physical activity; overuse syndrome

Citation: Djuricic, G.; Milanovic, F.; Ducic, S.; Radlović, V.; Lazovic, M.; Soldatovic, I.; Nikolic, D. Morphometric Parameters and MRI Morphological Changes of the Knee and Patella in Physically Active Adolescents. *Medicina* **2023**, *59*, 213. https://doi.org/10.3390/medicina59020213

Academic Editor: Johannes Mayr

Received: 3 December 2022
Revised: 16 January 2023
Accepted: 19 January 2023
Published: 22 January 2023

1. Introduction

Nowadays, the availability of a wide range of sports and the need for participation result in more children being physically active, particularly in competitive sports. The negative effect of improperly adapted exercise on the immature skeleton is well described in the literature [1–3]. However, an awareness of the potential unfavorable consequences is lacking. As a result, the incidence of sports injuries in the pediatric population due to unsuitable physical activity is increasing [4]. Repeated exercises might lead to overuse injuries of the overloaded joints and muscles due to the disproportion between musculoskeletal structures [5]. Physically active children are often referred to a doctor because of recurrent pain after excessive activity [6,7]. The most common consequence of those nonspecific, and at first glance insignificant, symptoms is an oversight of the hidden etiopathology [8]. The recurrence of such pain is rarely followed by any abnormal clinical and radiological findings, meaning patients might not be diagnosed in a timely fashion or might be inadequately treated.

Magnetic resonance imaging (MRI) is considered the reference standard modality in the radiological diagnosis of joints and soft tissue lesions. The high resolution offers excellent multiplanar imaging and differentiation of soft tissue, bone marrow, cartilage, muscles, ligaments, and tendons. Moreover, the absence of the negative effects of ionizing radiation makes MRI the method of choice to evaluate sports injuries in the pediatric population [9].

An analysis of the morphometric parameters of the patellofemoral joint and an assessment of their deviations from ideal values in the presence of patellofemoral dysplasia, as well as their relationship with the presence of risk factors for more frequent and extensive knee injuries in physically active children, were complemented by analyzing the patella type according to Wiberg, both for males and for females. Contrary to previous published works where [10,11] separately investigated these two risk factors, here we examined their joint impact and the mutual relationship between them. This is of particular importance both in the diagnosis and in the screening of sports-active children, with the aim of preventing injuries with a timely physiotherapy and conditioning program to improve knee biomechanics in people who are at the beginning of their sports career.

The main aim of this study is to present specific morphometric parameters and MRI morphological changes of the knee and patella in a physically active pediatric population. Additionally, we aimed to investigate the morphological risk factors for patellar instability.

2. Patients and Methods

In this prospective study, we observed the MRI findings of 193 physically active pediatric patients treated at the University Children's Hospital who reported pain in the knee joint during sports activities. Informed consent for participation was obtained from all subjects involved in the study.

Inclusion criteria were as follows: pediatric patients over 11 years old who participated for more than five and less than eight hours of sports activities per week during a period of more than three years of active training, followed by clinical examination and knee radiography in two views (AP and lateral view), without any pathological clinical and X-ray findings. Patients with osteomyelitis, bone tumors, juvenile rheumatoid arthritis, and metabolic diseases were excluded, as well as patients with previously treated trauma of the knee. The study was approved by the Institutional Review Board and followed the principles of good clinical practice (protocol number 017-16/15, 9 April 2021).

Two divisions of our study sample were performed: the first one considered the age of participants, and the second considered the MRI type of patella, according to Wiberg patellar classification. In the Wiberg classification, patellar facet sizes are classified into three types based on their relative medial and lateral sizes. The Wiberg classification initially applied to radiography but is now used in other modalities. [12]. Most patellae are Wiberg type 2, accounting for up to 65%. Type 3 accounts for 25%, and type 1 accounts for 10% [13]. In type 1 patella, the medial and lateral patellar facets are similar in size and are both concave; in type 2, the medial patellar facet is shorter than the lateral facet, and both facets are concave; in type 3 patella, the medial facet is much shorter than the lateral facet and convex [12]. Furthermore, a type 4 was described by Baumgartl, the "Jaegerhut" patella, with no medial facet and consequently no median ridge [12]. Age groups were set according to the American Academy of Pediatrics (AAP) classification; patients were divided into three age groups: group 1 (age 11–14 years), group 2 (age 15–17 years), and group 3 (age 18–21 years) [14]. MRI scans were performed on Siemens MagnetomAera 1.5T using a knee coil, positioned in supination with a field of view (FOV) of 16 cm using Ax Int FS, Cor Int FS, Cor T1, Sag Obl Int FS, Sag Obl PD, Cor Obl PD. The axial sections were parallel to the knee joint line; they included the entire patella and the head of the fibula. The coronal sections were parallel to the femoral condyle's posterior aspect—including the entire patella and up to 2 cm posterior to the femoral condyle. The sagittal sections were parallel to the medial aspect of the lateral condyle, including both collateral ligaments.

Magnetic resonance imaging was processed using Syngo via software to mark the following morphometric parameters: lateral trochlear inclination (LTI), trochlear facet asymmetry (TFA), the trochlear depth, the tibial tubercle–trochlear groove (TT-TG) distance, articular overlap, morphology ratio, and contact surface ratio.

The lateral trochlear inclination was determined from the first axial image in which cartilage was present and was used as a reference image. This angle corresponds to the angle between the lateral edge of the femoral epicondyles and the posterior edge of the femur. A line tangential to the subchondral bone of the posterior aspect of the two femoral condyles was crossed with a line tangential to the subchondral bone of the lateral trochlear facet to calculate LTI. A fat-suppressed PD-weighted image clearly delineates cartilage as a thick band of intermediate signal intensity adjacent to subchondral bone. When the inclination angle is less than 11°, trochlear dysplasia is diagnosed (sensitivity 93%; specificity 87%) [15] (Figure 1).

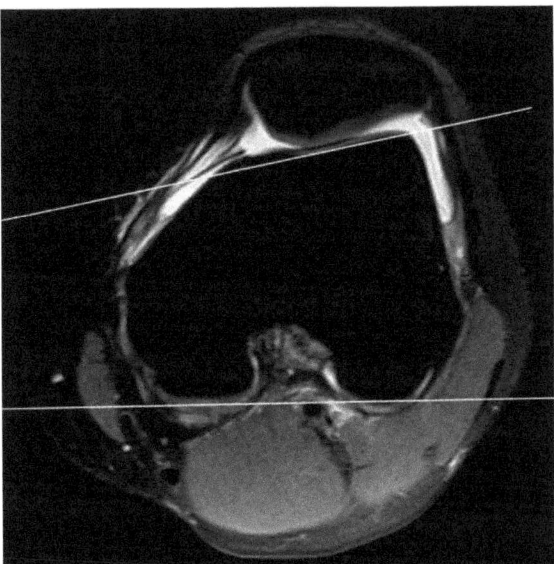

Figure 1. Proximal trochlear articular surface and measuring of lateral trochlear inclination.

The morphology of the trochlear facet asymmetry assessed on axial fat-saturated PD-weighted MR images, from the first axial image in which the articular cartilage was present. In this image, the greatest length and articular thickness of cartilage is approximately 3 cm above the tibiofemoral joint cleft. The asymmetry of the trochlear facet was measured as the ratio of the medial to lateral faces. In order to calculate asymmetry in the length of the medial facet (M) and the lateral facet (L), the medial facet length was divided by the lateral facet length and expressed as a percentage (M/L × 100%). In the presence of dysplasia, a trochlear facet ratio below 40% was considered indicative (96% specificity; 100% sensitivity) [16] (Figure 2).

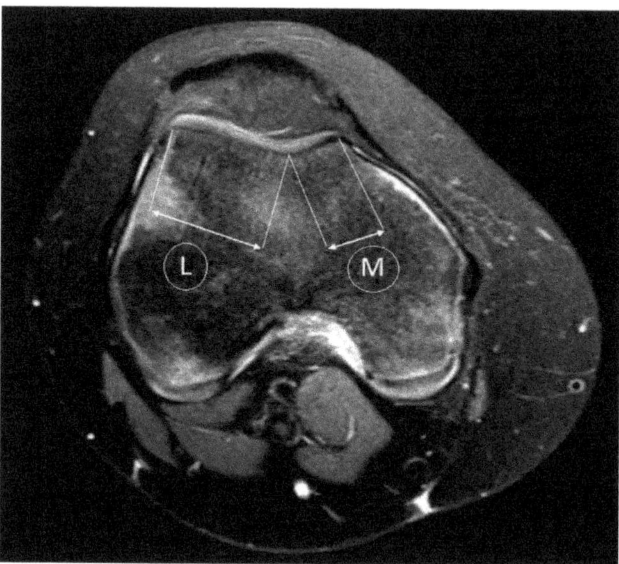

Figure 2. Trochlear facet asymmetry, ratio of length of medial trochlea to length of lateral trochlea. L—lateral, M—medial.

Using transverse sections, the trochlear depth was calculated using the formula ([A + B]/2 − C) given by Pfirmann et al. [17]. By measuring the maximum anteroposterior distance of the medial and lateral femoral condyles, distances A and B were determined. The distance C is the minimum anteroposterior distance between the deepest point of the trochlear groove and the line paralleling the posterior outlines of the femoral condyles.

Furthermore, the TT-TG measures the distance of the trochlear groove from the tibial tubercle. By using the posterior plane of the condyles as a reference line, the distance between the deepest point of the trochlea and the middle of the tibial tubercle was measured. In individuals with severe trochlear dysplasia, measuring the lateral distance between the tibial tubercle and the trochlear groove is less accurate because it is difficult to determine the depth of the trochlea. A distance of less than 15 mm between the tibial tubercle and the trochlear groove indicates normal tuberosity, a distance between 15 and 20 mm is borderline, and a distance longer than 20 mm indicates marked lateralization. It is nearly always associated with patellar instability if the distance between the tibial tubercle and trochlear groove exceeds 20 mm [16,18].

Additionally, we also observed patella alta ratios, and Insall–Salvati, modified Insall–Salvati, and Caton–Deschamps indices. Patella alta occurs when the patellar tendon is too long and rises too high above the trochlear fossa. Compared with a normal knee, patella alta requires a greater degree of flexion for the patella to engage in the trochlea, resulting in patellofemoral misalignment. In shallow degrees of flexion, this problem reduces the patellar contact area and decreases bone stability [19,20].

According to the Insall–Salvati ratio, patellar height is determined by the length of the patella tendon to the length of the patella. Sagital MR images are reliable tools for measuring patellar height ratios since they demonstrate the true anatomy of the patellofemoral joint and its ligaments in three dimensions. Distance lines for patellar tendon length (A), which represents the maximum length from the lower pole of the patella to its insertion on the tibia, and patellar length (B) are used in the calculation of Insall–Salvati ratio (A/B). An alternative method for assessing patellar height is the modified Insall–Salvati ratio. A proposal was made to avoid mismeasurements due to the variable shape of the inferior patella pole. The modified Insall–Salvati ratio (C/D) measures the distance from the inferior margin of the patellar articular surface (C), as opposed to the lower pole of the patella itself, to the patellar tendon insertion length of the patellar articular surface (D). Caton–Deschamps indices are based on the length of patellar articular surfaces and their distance from the tibia, which reduces erroneous measurements in patellas with long bodies, as measured by the Insall–Salvati ratio. The Caton–Deschamps index measures (A) the distance between the anterior angle of the tibial plateau and the most inferior aspect of the patellar articular surface length (B). Caton–Deschamps index = A/B. The Insall–Salvati ratio is considered normal for the values 0.8–1.2 (or values <2, according to MOD Insall–Salvati), whereas the Caton–Deschamps normal range is considered 0.6–1.3 [19].

The patellar trochlear overlap measures the articular overlap of the patellar undersurface and femoral trochlear cartilage (mm). The length of patellar cartilage (B) overlying the trochlear cartilage (A) was measured parallel to the subchondral surface of the patella using sagittal MRI. As an independent variable, the uncovered length of patellar cartilage (B) was also measured to calculate articular overlap as a percentage of overall articular length. Percentage coverage = (A/B) $\times$ 100 [21].

All measurements were performed independently by two observers, both with previous experience of more than 10 years in interpreting MRI findings, in order to evaluate interobserver reliability.

3. Statistical Analysis

The collected data were statistically analyzed using the IBM SPSS software package (IBM Corporation, New York NY, USA), version 21, using categorical and continuous variables. Descriptive statistics were used in order to determine the frequency of various pathological changes in knee joint injuries. A Pearson chi-square test, Fisher's exact test, and chi-square test for trend were performed. A statistical significance was considered at the level of $p < 0.05$. Continuous variables with a normal distribution were described using mean value and standard deviation (SD) to calculate the age range of patients.

4. Results

In our study, girls were frequently presented, 57%, versus 43% of boys. The age range was 11 to 21 years, with a mean of 15.8 ± 1.6. We found no statistical significance in age groups and side occurrence, as presented in Table 1.

Table 1. Knee injury prevalence by age groups.

		Age (Years)						
		11–14		**15–17**		**18–21**		*p*-Value
		N (%)	%	*N*	%	*N*	%	
Gender	male	22 (40.0)	40.0	48	44.4	13	43.3	0.863
	female	33 (60.0)	60.0	60	55.6	17	56.7	
Knee	Left	28 (50.9)	50.9	57	52.8	13	43.3	0.658
	right	27 (49.1)	49.1	51	47.2	17	56.7	

Age group 2 was the most frequent group, comprising more than half of the participants (56%), but the observed MRI parameters were not statistically significant in relation to age group, except the Caton–Deschamps index ($p = 0.039$), where the highest index value was noticed for age group 1 (1.13 ± 0.21) and the lowest for the age group 3 (1.03 ± 0.19). Even though we found no statistical significance for LTI ($p = 0.612$), TT-TG distance ($p = 0.801$), and Insall–Salvatti ($p = 0.127$) between age groups, the highest LTI was for age group 1 (15.75 ± 6.52), and the lowest was for the age group 3 (14.41 ± 5.89), while the TT-TG distance was highest for age group 1 (10.23 ± 5.01) and lowest for age group 2 (9.75 ± 4.34), and for Insall–Salvati, the highest index was for age group 1 (1.24 ± 0.21) and the lowest for age group 3 (1.14 ± 0.18). The data according to age groups are presented in Table 2.

Table 2. Morphometric parameters by age groups.

Parameter	Age Group (Years)	N	Mean ± SD	Median	*p*-Value
LTI	Group 1 (11–14)	55	15.75 ± 6.52	17.00	
	Group 2 (15–17)	108	15.40 ± 5.79	15.85	0.612
	Group 3 (18–21)	30	14.41 ± 5.89	15.50	
medial facet	Group 1 (11–14)	55	16.60 ± 2.78	16.60	
	Group 2 (15–17)	108	17.03 ± 3.49	17.00	0.357
	Group 3 (18–21)	30	17.65 ± 2.95	17.95	
lateral facet	Group 1 (11–14)	55	25.48 ± 3.72	25.00	
	Group 2 (15–17)	108	25.52 ± 3.95	25.50	0.861
	Group 3 (18–21)	30	25.92 ± 3.68	25.70	
trochlear facet asymmetry	Group 1 (11–14)	55	0.67 ± 0.14	0.69	
	Group 2 (15–17)	108	0.68 ± 0.16	0.69	0.707
	Group 3 (18–21)	30	0.69 ± 0.14	0.70	
TT-TG distance (mm)	Group 1 (11–14)	55	10.23 ± 5.01	10.10	
	Group 2 (15–17)	108	9.75 ± 4.34	10.23	0.801
	Group 3 (18–21)	30	9.81 ± 3.50	9.75	
Articular overlap	Group 1 (11–14)	55	12.70 ± 3.17	12.80	
	Group 2 (15–17)	108	13.13 ± 2.94	13.10	0.650
	Group 3 (18–21)	30	12.80 ± 2.58	12.55	
Insall–Salvati	Group 1 (11–14)	55	1.24 ± 0.21	1.23	
	Group 2 (15–17)	108	1.20 ± 0.22	1.18	0.127
	Group 3 (18–21)	30	1.14 ± 0.18	1.14	
MOD Insall–Salvati	Group 1 (11–14)	55	1.66 ± 0.23	1.64	
	Group 2 (15–17)	108	1.57 ± 0.23	1.56	0.062
	Group 3 (18–21)	30	1.56 ± 0.28	1.47	
Caton–Deschamps	Group 1 (11–14)	55	1.13 ± 0.21	1.10	
	Group 2 (15–17)	108	1.07 ± 0.16	1.05	0.039
	Group 3 (18–21)	30	1.03 ± 0.19	0.97	
Morphology ratio	Group 1 (11–14)	55	0.81 ± 0.08	0.82	
	Group 2 (15–17)	108	0.82 ± 0.08	0.83	0.463
	Group 3 (18–21)	30	0.82 ± 0.09	0.84	
Patellofemoral Contact Surface Ratio	Group 1 (11–14)	55	2.75 ± 0.73	2.55	
	Group 2 (15–17)	108	2.77 ± 0.86	2.52	0.979
	Group 3 (18–21)	30	2.77 ± 0.47	2.78	

Considering the patellar type in relation to gender and side affection, we found statistical significance related to gender ($p = 0.035$) but no statistical significance ($p = 0.826$) when considering side affection, as presented in Table 3. In age groups 1 and 2, females were more frequent (52.4% and 61.8%, respectively), while in the age group 3, males were more frequent (64.3%).

According to Wiberg morphological classification, the most frequent type was the second type, seen in about three quarters of tested individuals (74.6%), while the least frequent was the first patellar type, which was noticed in just above every tenth patient (10.9%) (Table 4).

Table 3. Knee injury prevalence by patellar type.

		Patellar Type						*p*-Value
		1		2		3		
		N	%	N	%	N	%	
Gender	male	10	47.6	55	38.2	18	64.3	0.035
	female	11	52.4	89	61.8	10	35.7	
Knee	Left	12	57.1	72	50.0	14	50.0	0.826
	right	9	42.9	72	50.0	14	50.0	

Table 4. Morphometric parameters by patellar type (Wiberg classification).

Parameter Patellar Type		N	Mean ($\pm$ SD)	Median	*p*-Value
LTI	1	21	14.79 $\pm$ 4.94	16.10	<0.001
	2	144	16.29 $\pm$ 5.63	16.80	
	3	28	10.89 $\pm$ 6.72	9.25	
medial facet	1	21	19.17 $\pm$ 2.58	19.50	<0.001
	2	144	17.33 $\pm$ 2.95	17.40	
	3	28	13.69 $\pm$ 2.66	13.70	
lateral facet	1	21	22.59 $\pm$ 2.95	22.20	<0.001
	2	144	25.13 $\pm$ 3.25	25.25	
	3	28	30.09 $\pm$ 3.54	30.00	
trochlear facet asymmetry	1	21	0.85 $\pm$ 0.10	0.87	<0.001
	2	144	0.69 $\pm$ 0.11	0.70	
	3	28	0.46 $\pm$ 0.11	0.46	
TT-TG distance (mm)	1	21	9.20 $\pm$ 3.98	10.20	0.086
	2	144	10.29 $\pm$ 4.40	10.05	
	3	28	8.40 $\pm$ 4.51	7.77	
Articular overlap	1	21	13.15 $\pm$ 3.35	13.10	0.952
	2	144	12.93 $\pm$ 2.70	13.00	
	3	28	12.94 $\pm$ 3.85	13.15	
Insall–Salvati	1	21	1.22 $\pm$ 0.23	1.18	0.001
	2	144	1.17 $\pm$ 0.20	1.15	
	3	28	1.33 $\pm$ 0.21	1.36	
MOD Insall–Salvati	1	21	1.60 $\pm$ 0.19	1.59	0.834
	2	144	1.59 $\pm$ 0.25	1.56	
	3	28	1.62 $\pm$ 0.22	1.61	
Caton–Deschamps	1	21	1.07 $\pm$ 0.13	1.04	0.018
	2	144	1.06 $\pm$ 0.19	1.03	
	3	28	1.17 $\pm$ 0.16	1.14	
Morphology ratio	1	21	0.82 $\pm$ 0.07	0.82	0.059
	2	144	0.81 $\pm$ 0.08	0.83	
	3	28	0.85 $\pm$ 0.06	0.85	
PF Contact Surface Ratio	1	21	2.76 $\pm$ 0.87	2.48	0.308
	2	144	2.73 $\pm$ 0.67	2.56	
	3	28	2.97 $\pm$ 1.10	2.53	

The average LTI in different types of the patella was 14.79 ± 4.94 for the first patella type according to Wiberg, 16.29 ± 5.63 for the second, and 10.89 ± 6.72 for the third patellar type. The values are expressed in Table 4. According to our study, we found 50 participants with LTI < 11°, and that the patellar types correlate with LTI < 11°. In patellar type 1, 6 (28.6%) participants had LTI < 11°, compared to 28 (19.4%) in type 2 and 16 (57.1%) in type 3 ($p < 0.001$). For comparing LTI < 11° and patellar type, the Pearson chi-square test was used.

Furthermore, there was a statistically significant association between patellar type groups and trochlear facet asymmetry ($p < 0.001$), Insall–Salvati index (Figure 3) ($p = 0.001$), and Caton–Deschamps index (Figure 4) ($p = 0.018$) (Table 4). Regarding TFA, the highest values were for patellar type 1 according to Wiberg (0.85 ± 0.10), and the lowest were for patellar type 3 according to Wiberg (0.46 ± 0.11). For the Insall–Salvati index, the highest values were for patellar type 3 according to Wiberg (1.33 ± 0.21), and the lowest were for patellar type 2 according to Wiberg (1.17 ± 0.20). For the Caton–Deschamps index, the highest values were for patellar type 3 according to Wiberg (1.17 ± 0.16), and the lowest were for patellar type 2 according to Wiberg (1.06 ± 0.19).

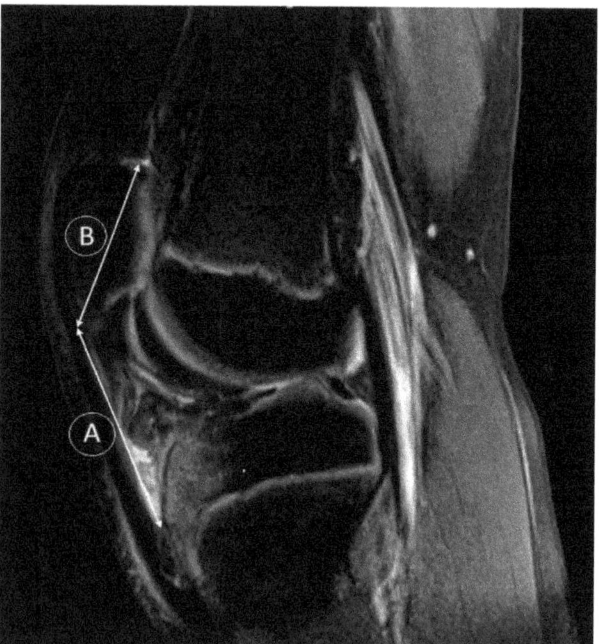

Figure 3. Patellar Insall–Salvati index (A:B). A—patellar tendon length, B—patellar length (described in the section Materials and Methods).

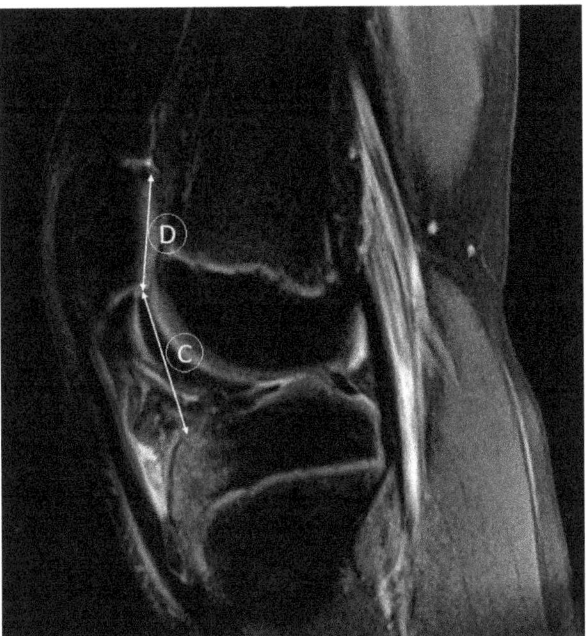

Figure 4. Patellar Caton–Deschamps index (C:D). C—distance from the inferior margin of the patellar articular surface, D—patellar tendon insertion length of the patellar articular surface (described in the section Materials and Methods)

5. Discussion

In our study, we have demonstrated that LTI, TFA, Insall–Salvati index, and Caton–Deschamps index significantly differed with regards to the patellar type according to Wiberg. Furthermore, only the Caton–Deschamps index, among the studied parameters, significantly differed with regards to the age groups of tested participants.

The prediction of children's injury rate using various morphological parameters of the knee has always been of interest and a debatable topic. In our study, it was shown that certain types of patella (according to Wiberg classification) might increase susceptibility for a knee injury, depending on the age of participants [14]. Furthermore, it was shown that the most common type of patella in knee injury (according to Wiberg) is type 2 in boys and girls type 3.

Trochlear dysplasia (TD) is one of the leading causes of patellar instability [22]. The most commonly used MRI parameters for TD are LTI and TFA [23]. The LTI is commonly used to distinguish physiological from potentially dysplastic knees. LTI provides a quantitative description of dysplasia; it is reported that an LTI below 11° is associated with a 95% specificity of having patellar instability secondary to trochlear asymmetry [24], while a threshold of <0.4 or 40% is suggestive of trochlear dysplasia [25]. According to our study, the most vulnerable group of examined patients for TD was the one with the patellar type 2 due to the Wiberg classification. Carrillon et al. set the cut-off of 11 degrees for trochlear dysplasia, with very high sensitivity [15]. Stepanovich et al. explained that the existing growth potential affects the threshold, so the authors suggested the lower limit of 17 instead of 11 degrees [26]. Our results stress that the threshold for LTI might be at 11 degrees. Moreover, our findings indicate that certain patella types could be associated with increased susceptibility for TD in children after knee injury. Finally, we found that during skeletal maturation, LTI decreased, and TFA increased in a certain type of patella in knee injury in children. Moreover, the TFA ratio reported in our study was somewhat higher than that previously reported [27,28]. The possible explanation for our findings and the discrepancy compared to previous reports might be the fact that our study sample included sport-active adolescents. It was noticed that cartilage thickness at the patella does not strongly correlate with age [29], while physical activity might influence the thickness of articular cartilage [30].

The TT-TG distance value is also an essential parameter for evaluating patellofemoral disorders [31]. According to the literature, the LTI and TT-TG distance are used for diagnostics, treatment planning, and predicting future risk of re-dislocations [25]. Stepanovich et al. reported the correlation between TT-TG distance and patellar instability in patellar dislocation. They showed pathological values for TT-TG distance (over 20 mm) on their sample [26], which was not reflected in the results that we obtained. Our study found no statistical significance in TT-TG values in different patella types in different age groups. This might be explained by the differences in study design in regard to the observed groups. We observed participants with no injury history and with an absence of pathological clinical and X-ray findings, contrary to Stepanovich and co-workers. Additionally, the wide range of participants in different age groups in our study could be one of possible factors influencing the statistical outcome for the TT-TG parameter. Therefore, further study should include more homogenous age-group samples.

The position of the patella has an essential role in both knee stability and knee instability. The most frequently used measures in patella alta ratios are the Insall–Salvati, modified Insall–Salvati, and Caton–Deschamps indices [32,33]. We found statistical significance in the Caton–Deschamps index when considering the age groups of participants. According to different types of the patella (according to Wiberg), we found statistical significance of the Insall–Salvati index and the Caton–Deschamps index. The values of Insall–Salvati ratio in our study correspond with those of previous reports [34]. The same applies for the Caton–Deschamps index in our study [35]. Moreover, our study suggested that the Caton–Deschamps index is somehow more age-sensitive, because the knee cartilage changes described by this index were more detectable compared to the morphological changes described by the above-mentioned parameters.

Furthermore, we showed that patella type 2 in boys and patella type 3 in girls (according to Wiberg) are the most prone to a knee injury in physically active children.

The comparisons of the present study results with those of the current literature are presented in Supplementary Material 1.

6. Conclusions

According to our study, the most frequent knee injury parameter in physically active pediatric patients is Wiberg patella type 2 in boys and type 3 in girls. Those patellar type morphological parameters, in association with high patella alta index, expressed as Insall–Salvati index and Caton–Deschamps index, could influence knee injury susceptibility in physically active children, especially in children over 14 years.

The MRI morphometric parameters observed in our study might be factors of prediction of knee injury in physically active children. Moreover, it might be very useful in sports programs to improve the biomechanics of the knee in order to reduce the injury rate in sports-active children.

Supplementary Materials: The following supporting information can be downloaded at: https://www.mdpi.com/article/10.3390/medicina59020213/s1, Table S1: material 1.

Author Contributions: G.D., V.R., D.N., and F.M.—conceptualization, investigation, methodology, supervision, aFnd writing original draft; S.D., I.S., and M.L.—formal analysis, data curation, and writing original draft. All authors have read and agreed to the published version of the manuscript.

Funding: The authors report no involvement in the research by the sponsor that could have influenced the outcome of this work.

Institutional Review Board Statement: The study was approved by the Institutional Review Board and followed the principles of good clinical practice (protocol number 017-16/15, 9 April 2021).

Informed Consent Statement: Informed consent was obtained from all subjects involved in the study.

Data Availability Statement: All the data are available from the corresponding author upon reasonable request.

Conflicts of Interest: The authors certify that there is no conflict of interest with any financial organization regarding the material discussed in the manuscript.

References

1. Sweeney, E.; Rodenberg, R.; MacDonald, J. Overuse Knee Pain in the Pediatric and Adolescent Athlete. *Curr. Sport. Med. Rep.* **2020**, *19*, 479–485. [CrossRef] [PubMed]
2. DiFiori, J.P.; Benjamin, H.J.; Brenner, J.S.; Gregory, A.; Jayanthi, N.; Landry, G.L.; Luke, A. Overuse injuries and burnout in youth sports: A position statement from the American Medical Society for Sports Medicine. *Br. J. Sport. Med.* **2014**, *48*, 287–288. [CrossRef] [PubMed]
3. Caine, D.; Meyers, R.; Nguyen, J.; Schoffl, V.; Maffulli, N. Primary Periphyseal Stress Injuries in Young Athletes: A Systematic Review. *Sport. Med.* **2021**, *52*, 741–772. [CrossRef]
4. Arnold, A.; Thigpen, C.A.; Beattie, P.F.; Kissenberth, M.J.; Shanley, E. Overuse Physeal Injuries in Youth Athletes. *Sport. Health* **2017**, *9*, 139–147. [CrossRef] [PubMed]
5. Bateni, C.P.; Bindra, J.; Haus, B. MRI of Sports Injuries in Children and Adolescents: What s Different from Adults. *Curr. Radiol. Rep.* **2014**, *2*, 45. [CrossRef]
6. Yen, Y.M. Assessment and treatment of knee pain in the child and adolescent athlete. *Pediatr. Clin. N. Am.* **2014**, *61*, 1155–1173. [CrossRef]
7. Patel, D.R.; Villalobos, A. Evaluation and management of knee pain in young athletes: Overuse injuries of the knee. *Transl. Pediatr.* **2017**, *6*, 190–198. [CrossRef]
8. Barton, C.J.; Lack, S.; Hemmings, S.; Tufail, S.; Morrissey, D. The 'Best Practice Guide to Conservative Management of Patellofemoral Pain': Incorporating level 1 evidence with expert clinical reasoning. *Br. J. Sport. Med.* **2015**, *49*, 923–934. [CrossRef]
9. Pooley, R.A. AAPM/RSNA physics tutorial for residents: Fundamental physics of MR imaging. *Radiographics* **2005**, *25*, 1087–1099. [CrossRef]
10. Gudas, R.; Šiupšinskas, L.; Gudaitė, A.; Vansevičius, V.; Stankevičius, E.; Smailys, A.; Vilkytė, A.; Simonaitytė, R. The Patello-Femoral Joint Degeneration and the Shape of the Patella in the Population Needing an Arthroscopic Procedure. *Medicina* **2018**, *54*, 21. [CrossRef]
11. Zaffagnini, S.; Grassi, A.; Zocco, G.; Rosa, M.A.; Signorelli, C.; Marcheggiani Muccioli, G.M. The patellofemoral joint: From dysplasia to dislocation. *EFORT Open Rev.* **2017**, *2*, 204–214. [CrossRef]
12. Meyers, A.B.; Laor, T.; Sharafinski, M.; Zbojniewicz, A.M. Imaging assessment of patellar instability and its treatment in children and adolescents. *Pediatr. Radiol.* **2016**, *46*, 618–636. [CrossRef]
13. Stoller, D.W.; Sampson, T.G.; Li, A.E. The knee. In *Magnetic Resonance Imaging in Orthopaedics and Sports Medicine*, 3rd ed.; Stoller, D.W., Ed.; Lippincott Williams & Wilkins: Philadelphia, PA, USA, 2007; pp. 305–732.
14. Available online: https://www.healthychildren.org/English/ages-stages/teen/Pages/Stages-of-Adolescence.aspx (accessed on 5 June 2022).
15. Carrillon, Y.; Abidi, H.; Dejour, D.; Fantino, O.; Moyen, B.; Tran-Minh, V.A. Patellar instability: Assessment on MR images by measuring the lateral trochlear inclination-initial experience. *Radiology* **2000**, *216*, 582–585. [CrossRef] [PubMed]
16. Diederichs, G.; Issever, A.S.; Scheffler, S. MR imaging of patellar instability: Injury patterns and assessment of risk factors. *Radiographics* **2010**, *30*, 961–981. [CrossRef] [PubMed]
17. Pfirrmann, C.W.; Zanetti, M.; Romero, J.; Hodler, J. Femoral trochlear dysplasia: MR findings. *Radiology* **2000**, *216*, 858–864. [CrossRef] [PubMed]
18. Wilcox, J.J.; Snow, B.J.; Aoki, S.K.; Hung, M.; Burks, R.T. Does landmark selection affect the reliability of tibial tubercle-trochlear groove measurements using MRI? *Clin. Orthop. Relat. Res.* **2012**, *470*, 2253–2260. [CrossRef]
19. Charles, M.D.; Haloman, S.; Chen, L.; Ward, S.R.; Fithian, D.; Afra, R. Magnetic resonance imaging-based topographical differences between control and recurrent patellofemoral instability patients. *Am. J. Sport. Med.* **2013**, *41*, 374–384. [CrossRef]
20. Dejour, H.; Walch, G.; Nove-Josserand, L.; Guier, C. Factors of patellar instability: An anatomic radiographic study. *Knee Surg. Sport. Traumatol. Arthrosc.* **1994**, *2*, 19–26. [CrossRef]
21. Munch, J.L.; Sullivan, J.P.; Nguyen, J.T.; Mintz, D.; Green, D.W.; Shubin Stein, B.E.; Strickland, S. Patellar Articular Overlap on MRI Is a Simple Alternative to Conventional Measurements of Patellar Height. *Orthop. J. Sport. Med.* **2016**, *4*, 2325967116656328. [CrossRef]
22. Trasolini, N.A.; Serino, J.; Dandu, N.; Yanke, A.B. Treatment of Proximal Trochlear Dysplasia in the Setting of Patellar Instability: An Arthroscopic Technique. *Arthrosc. Tech.* **2021**, *10*, e2253–e2258. [CrossRef]

23. Steensen, R.N.; Bentley, J.C.; Trinh, T.Q.; Backes, J.R.; Wiltfong, R.E. The prevalence and combined prevalences of anatomic factors associated with recurrent patellar dislocation: A magnetic resonance imaging study. *Am. J. Sport. Med.* **2015**, *43*, 921–927. [CrossRef] [PubMed]

24. Joseph, S.M.; Cheng, C.; Solomito, M.J.; Pace, J.L. Lateral Trochlear Inclination Angle: Measurement via a 2-Image Technique to Reliably Characterize and Quantify Trochlear Dysplasia. *Orthop. J. Sport. Med.* **2020**, *8*, 2325967120958415. [CrossRef]

25. Paiva, M.; Blond, L.; Holmich, P.; Steensen, R.N.; Diederichs, G.; Feller, J.A.; Barfod, K.W. Quality assessment of radiological measurements of trochlear dysplasia; a literature review. *Knee Surg. Sport. Traumatol. Arthrosc.* **2018**, *26*, 746–755. [CrossRef] [PubMed]

26. Stepanovich, M.; Bomar, J.D.; Pennock, A.T. Are the Current Classifications and Radiographic Measurements for Trochlear Dysplasia Appropriate in the Skeletally Immature Patient? *Orthop. J. Sport. Med.* **2016**, *4*, 2325967116669490. [CrossRef]

27. Shen, J.; Qin, L.; Yao, W.W.; Li, M. The significance of magnetic resonance imaging in severe femoral trochlear dysplasia assessment. *Exp. Ther. Med.* **2017**, *14*, 5438–5444. [CrossRef]

28. Kim, H.K.; Parikh, S. Patellofemoral Instability in Children: Imaging Findings and Therapeutic Approaches. *Korean J. Radiol.* **2022**, *23*, 674–687. [CrossRef]

29. Sidharthan, S.; Yau, A.; Almeida, B.A.; Shea, K.G.; Greditzer, H.G.; Jones, K.J.; Fabricant, P.D. Patterns of Articular Cartilage Thickness in Pediatric and Adolescent Knees: A Magnetic Resonance Imaging-Based Study. *Arthrosc. Sport. Med. Rehabil.* **2021**, *3*, e381–e390. [CrossRef]

30. Hori, M.; Terada, M.; Suga, T.; Isaka, T. Changes in anterior femoral articular cartilage structure in collegiate rugby athletes with and without a history of traumatic knee joint injury following a five-month competitive season. *Sci. Rep.* **2021**, *11*, 15186. [CrossRef]

31. Nagamine, R.; Miura, H.; Inoue, Y.; Tanaka, K.; Urabe, K.; Okamoto, Y.; Nishizawa, M.; Iwamoto, Y. Malposition of the tibial tubercle during flexion in knees with patellofemoral arthritis. *Skelet. Radiol.* **1997**, *26*, 597–601. [CrossRef]

32. Verhulst, F.V.; van Sambeeck, J.D.P.; Olthuis, G.S.; van der Ree, J.; Koeter, S. Patellar height measurements: Insall-Salvati ratio is most reliable method. *Knee Surg. Sport. Traumatol. Arthrosc.* **2020**, *28*, 869–875. [CrossRef] [PubMed]

33. Cavallo, F.; Mohn, A.; Chiarelli, F.; Giannini, C. Evaluation of Bone Age in Children: A Mini-Review. *Front. Pediatr.* **2021**, *9*, 580314. [CrossRef] [PubMed]

34. Kurowecki, D.; Shergill, R.; Cunningham, K.M.; Peterson, D.C.; Takrouri, H.S.R.; Habib, N.O.; Ainsworth, K.E. A comparison of sagittal MRI and lateral radiography in determining the Insall-Salvati ratio and diagnosing patella alta in the pediatric knee. *Pediatr. Radiol.* **2022**, *52*, 527–532. [CrossRef] [PubMed]

35. Thévenin-Lemoine, C.; Ferrand, M.; Courvoisier, A.; Damsin, J.P.; Ducou le Pointe, H.; Vialle, R. Is the Caton-Deschamps index a valuable ratio to investigate patellar height in children? *J. Bone Jt. Surg.* **2011**, *93*, e35. [CrossRef] [PubMed]

Article

Hemp Seed Oil in Association with β-Caryophyllene, Myrcene and Ginger Extract as a Nutraceutical Integration in Knee Osteoarthritis: A Double-Blind Prospective Case-Control Study

Giacomo Farì [1,2,*], Marisa Megna [1], Salvatore Scacco [1,3,*], Maurizio Ranieri [1], Maria Vittoria Raele [1], Enrica Chiaia Noya [1], Dario Macchiarola [1,4], Francesco Paolo Bianchi [5], Davide Carati [6], Simona Panico [6], Eleonora Di Campi [6], Antonio Gnoni [1], Venera Scacco [1], Alessio Danilo Inchingolo [7], Erda Qorri [8], Antonio Scarano [9] and Biagio Rapone [7]

[1] Department of Translational Biomedicine and Neuroscience (DiBraiN), Aldo Moro University, 70121 Bari, Italy
[2] Department of Biological and Environmental Science and Technologies (Di.S.Te.B.A.), University of Salento, 73100 Lecce, Italy
[3] Mater Dei Hospital C.B.H., 70125 Bari, Italy
[4] Istituti Clinici Scientifici Maugeri—IRCCS Bari, 70124 Bari, Italy
[5] Department of Biomedical Science and Human Oncology, Aldo Moro University of Bari, 70121 Bari, Italy
[6] Ansce Bio Generic, 73020 Carpignano Salentinono, Italy
[7] Interdisciplinary Department of Medicine, University of Bari "Aldo Moro", 70121 Bari, Italy
[8] Dean Faculty of Medical Sciences, Albanian University, 1001 Tirana, Albania
[9] Department of Oral Science, Nano and Biotechnology, CaSt-Met University of Chieti-Pescara, 66100 Chieti, Italy
[*] Correspondence: giacomo.fari@unisalento.it (G.F.); salvatore.scacco@uniba.it (S.S.); Tel.: +39-0836575410 (G.F.); +39-0805448514 (S.S.)

Citation: Farì, G.; Megna, M.; Scacco, S.; Ranieri, M.; Raele, M.V.; Chiaia Noya, E.; Macchiarola, D.; Bianchi, F.P.; Carati, D.; Panico, S.; et al. Hemp Seed Oil in Association with β-Caryophyllene, Myrcene and Ginger Extract as a Nutraceutical Integration in Knee Osteoarthritis: A Double-Blind Prospective Case-Control Study. *Medicina* **2023**, *59*, 191. https://doi.org/10.3390/medicina59020191

Academic Editors: Dejan Nikolić and Milica Lazovic

Received: 22 December 2022
Revised: 11 January 2023
Accepted: 15 January 2023
Published: 18 January 2023

Abstract: *Background and Objectives:* Nutraceuticals are gaining more and more importance as a knee osteoarthritis (KOA) complementary treatment. Among nutraceuticals, hemp seed oil and terpenes are proving to be very useful as therapeutic support for many chronic diseases, but there are still few studies regarding their effectiveness for treating KOA, both in combination and separately. The aim of this study is thus to compare the effect of two dietary supplements, both containing hemp seed oil, but of which only one also contains terpenes, in relieving pain and improving joint function in patients suffering from KOA. *Materials and Methods:* Thirty-eight patients were recruited and divided into two groups. The control group underwent a 45 day treatment with a hemp seed oil-based dietary supplement, while the treatment group assumed a hemp seed oil and terpenes dietary supplement for the same period. Patients were evaluated at the enrollment (T0) and at the end of treatment (T1). Outcome measures were: Numeric Rating Scale (NRS), Oswestry Disability Index (ODI), Short-Form-12 (SF-12), Knee Injury Osteoarthritis Outcome Score (KOOS), and Oxford Knee Score (OKS). *Results:* All outcome measures improved at T1 in both groups, but NRS, KOOS and OKS had a greater significant improvement in the treatment group only. *Conclusions:* Hemp seed oil and terpenes resulted a more effective integrative treatment option in KOA, improving joint pain and function and representing a good complementary option for patients suffering from osteoarthritis.

Keywords: knee osteoarthritis; rehabilitation; dietary supplement; hemp seed oil; terpenes

1. Introduction

Knee osteoarthritis (KOA) is the most common form of limb osteoarthritis [1]. It is a chronic joint disease which is characterized by degenerative lesions of the articular cartilage that progressively cause pain, motor impairment and, in the most severe cases, deformation of the joint itself [2,3]. Although one of the main risk factors is old age, as KOA predominantly affects people in their 70s [4], KOA is also increasing in patients aged between 45 and 70 [5], emerging recently as a very relevant problem for society and placing it among the most common causes of disability [6]. At present, KOA remains an untreatable condition because its mechanisms of progression are not fully understood [7].

Therefore, the goal of osteoarthritis treatment is to alleviate symptoms and to slow down the disease progression. The KOA therapeutic spectrum ranges from pharmacotherapy to physical therapies, orthotics and, finally, surgery and rehabilitation [8–12]. In recent years, nutraceuticals, which are dietary supplements used to improve health, delay aging, prevent disease, and support the functioning of the human body, are gaining importance [13]. In patients with osteoarthritis, the assumption of long-chain omega-3 fattyacids from fish oil supplements and micronutrients such as vitamin K is considered very useful [14], since it has a role in bone and cartilage mineralization. New molecules are constantly emerging, including components of industrial hemp (Cannabis sativa, Cannabacae), which have already been shown to be effective for anxiety disorders and in reducing oxidative stress, contrasting the risk of chronic diseases including joint diseases, neurological disorders, digestive problems and skin conditions [15,16]. Recent studies showed important preclinical and clinical evidence about Cannabis Sativa pain relief properties [17,18], especially because it contains two main phytocannabinoids: D9-tetrahydrocannabinol (D9-THC) and cannabidiol (CBD) [19,20].

Among the several components of Cannabis Sativa, as flavonoids, vitamins, fatty acids, sterols, lignanamides, spiroindans, and alkaloids that may have health benefits [21], terpenes represent a very attractive option for pain treatment [22]. In particular, β-caryophyllene (BCP), a bicyclic sesquiterpene very present in Cannabis Sativa, has been widely investigated and highly appreciated for its low toxicity and considerable safety profile [23]. One of its main targets was described to be the cannabinoid receptor type 2 (CB2 receptor), for which it is thought to act as a full agonist [24]. Interestingly, recent data suggest that the selective agonism of CB2 receptors may constitute a novel strategy for treating chronic pain [25]. β-Myrcene is a monoterpene composed of two isoprene units, and a recent study showed its significant anti-inflammatory and anti-catabolic effects in human chondrocytes and, thus, its ability to halt or, at least, slow down cartilage destruction and osteoarthritis progression [26]. Ginger extract takes advantage of the anti-inflammatory properties of gingerols and shagaols, which selectively inhibit COX-2 (cyclooxygenase-2) and inflammatory cytokines [27]. Nevertheless, there remains little evidence with regard to the effects of Cannabis Sativa components on osteoarthritis pain management, especially from a clinical point of view.

Thus, the aim of this study is to compare the effect of two regimens of food supplementation in relieving pain and to improve joint function in patients aged between 45 and 70 suffering from KOA: the first one based exclusively on hemp seed oil (without cannabinoids), the second one based on hemp seed oil (without cannabinoids) but potentiated with terpenes (β-caryophyllene and myrcene). This comparison could be useful to understand if the two nutraceuticals are useful and if one of these is better than the other, considering their different composition.

2. Materials and Methods

The study model is a double-blind prospective case-control study. The study was carried out in the period between March and August 2022. Patients were enrolled if they met the following criteria: age between 45 and 70 years; a clinical diagnosis of KOA according to the American College of Rheumatology criteria; knee pain $\geq$ 4 according to the Numeric Rating Scale (NRS) at the enrollment and in the previous 15 days; radiographic KOA classifiable as grade II-III according to Kellgren-Lawrence scale; ability to understand the purpose and design of the study, and to provide informed consent. Exclusion criteria were: KOA local complications (e.g., hematoma and joint effusion); knee pain due to trauma (during the previous month); any disease potentially interfering with medical evaluation different from KOA (e.g., rheumatoid arthritis, metabolic inflammatory arthropathy); local drug infiltration (hyaluronic acid, steroids, stem cells, polynucleotides, Platelet Rich Plasma) or physiotherapy (e.g., laser therapy, shock wave therapy, therapeutic exercise, etc.) within the previous 45 days; assumption of non-steroidal anti-inflammatory drugs or analgesics within 15 days prior the enrollment; assumption of slow-acting drugs or dietary supplements in the previous 3 months (e.g., chondroitin sulfate, diacerein, soybean and avocado unsaponifiables, oxaceprol, granions de cuivre, glucosamine, phytotherapy for osteoarthritis); contraindications to acetaminophen; systemic diseases which contraindicate nutraceuticals assumption (liver failure, kidney failure, uncontrolled cardiovascular disease); pregnant or lactating women; pre-menopausal women not using contraception; and patients enrolled in other clinical trials within the past three months.

Thirty-eight patients with monolater KOA were recruited and then divided into two groups, each consisting of nineteen subjects.

At the time of recruitment (T0), all patients underwent a medical examination, which included medical history, standardized physical examination, and x-rays evaluation. Therefore, the weight and height of each patient were detected and the Body Mass Index (BMI) was calculated according to the formula: weight (Kg)/height (m^2). The following rating scales were then measure for each patient:

- Numeric Rating Scale (NRS): this is a one-dimensional scale that rates pain from 0, the absence of pain, to 10, the maximum perceived pain;
- Oswestry Disability Index (ODI): this is a scale that rates the percentage value of disability and ranges from 0%, no disability, to 100%, maximum disability;
- Short Form 12 (SF-12): this is a quality of life assessment scale. It is divided into physical domain (PCS) and mental domain (MCS). The higher the score, the better the patients' quality of life;
- Knee Injury and Osteoarthritis Outcome Score (KOOS): this is a percentage value that quantifies clinical symptoms, disability, and quality of life in patients suffering from knee diseases. It ranges from 0% (severe disability) to 100% (optimal condition);
- Oxford Knee Score (OKS): this assesses the severity of osteoarthritis from 0 to 48 (severe osteoarthritis 0–19; moderate-severe osteoarthritis 20–29; mild-moderate osteoarthritis 30–39; no sign of osteoarthritis 40–48).

Patients belonging to the control group underwent a 45 day treatment with a dietary supplement based on hemp seed oil (413 mg/capsule) in a softgel capsule format.

Patients belonging to the treatment group underwent a 45 day treatment with a dietary supplement based on hemp seed oil (413 mg/capsule), β-caryophyllene (35 mg/capsule), myrcene (15 mg/capsule), and ginger extract titrated in gingerols (66 mg/capsule). The hemp seed oil contained in both dietary supplements was composed mainly of Linoleic (55.90%), gamma-Linolenic (19.10%) and Oleic (9.30%) acids. Patients were unaware of which of the two dietary supplements they were taking as they were not identifiable from the packaging. Similarly, the physicians performing the clinical assessments at T0 and T1 were unaware of which supplement the patients had taken, thus creating a double-blind study design. A third investigator was therefore responsible for the distribution of the supplements.

All patients took two softgels of the assigned dietary supplement per day, one capsule with each main meal (usually during lunch and dinner).

Patients were allowed to take paracetamol (up to a maximum of 3000 mg/day) and were asked to write down the dosage of the drug taken in a dedicated diary. The use of other medications during the treatment period was recorded, as were the eventual side effects.

At the end of the treatment (T1), 45 days after T0, all of the outcome measures were collected from each patient in order to compare the clinical trend and the functional implications between the groups. The diaries used to register any drugs taken in addition to any side effects related to the proposed therapies were collected at the same time.

All patients received the necessary information during the first medical examination and expressed their written informed consent. All of the performed procedures were carried out in accordance with the Helsinki Declaration (2016) of the World Medical Association. The study was approved by the Ethics Committee of Albania University, Tiran, Albania (Nr. 587 Prot.–Date: 13 December 2021).

Statistical Analysis

A data analysis was performed using STATA MP17 software. Continuous variables were described as mean $\pm$ standard deviation (SD) and range, and categorical variables as proportions. A skewness and kurtosis test was used to evaluate the normality of continuous variables and a normalization model was constructed using the logarithmic function for those not normally distributed. The Student's *t*-test for independent data was used to compare continuous variables between groups, and the ANOVA for repeated measures test was used to compare continuous variables between groups and detection time. Multivariate linear regression was used to assess the relationship between the difference from T1 to T0 of each individual outcome and the group (treatment vs. control), sex (male vs. female), age (years) and BMI; correlation coefficients were calculated, with a 95% confidence interval (95%CI) indicated. A *p*-value < 0.05 was considered significant for all tests.

3. Results

The study sample was made up of 38 subjects, of which 19 (50.0%) belonged to the control group and 19 (50.0%) belonged to the treatment group; the characteristics of the sample, by group, are shown in Table 1.

Table 1. Characteristics of the sample, by group.

Variable	Control (*n* = 19)	Treatment (*n* = 19)	Total (*n* = 38)	*p*-Value
Females; *n* (%)	10 (52.6)	10 (52.6)	20 (52.6)	1.000
Age(years); mean $\pm$ SD (range)	59.7 $\pm$ 6.6 (47–69)	54.5 $\pm$ 4.6 (48–65)	57.1 $\pm$ 6.2 (47–69)	0.008
BMI; mean $\pm$ SD (range)	27.9 $\pm$ 3.8 (20.4–33.4)	29.6 $\pm$ 6.3 (20.0–49.3)	28.7 $\pm$ 5.2 (20.0–49.3)	0.376

Control = control group; treatment = treatment group; BMI = Body Mass Index; SD = Standard Deviation; *n* = number.

The outcome variables, by group and detection time, are described in Table 2 and Figures 1–6; the ANOVA test for repeated measures showed a statistically significant difference for all the outcome measures in the comparison between T0 and T1 ($p < 0.0001$). The same test showed a statistically significant difference for NRS, KOOS and OKS scores in the interaction between T0 and T1 and between the two groups ($p < 0.0001$). All of these findings are described in Table 2.

Table 2. Mean $\pm$ SD and range of outcome variables, by group and detection time.

Variable	Control (*n* = 19)	Treatment (*n* = 19)	Total (*n* = 38)	Group Comparison	Time Comparison	Time and Group Interaction
NRS T0	7.6 $\pm$ 1.4 (6–10)	8.3 $\pm$ 1.1 (7–10)	7.9 $\pm$ 1.3 (6–10)	0.080	<0.0001	<0.0001
NRS T1	5.7 $\pm$ 1.2 (4–8)	3.5 $\pm$ 2.1 (1–8)	4.6 $\pm$ 2.0 (1–8)			
ODI T0	31.3 $\pm$ 18.9 (6–72)	29.8 $\pm$ 12.8 (10–64)	30.6 $\pm$ 15.9 (6–72)	0.687	<0.0001	0.963
ODI T1	17.2 $\pm$ 9.3 (6–44)	15.9 $\pm$ 7.2 (6–34)	16.6 $\pm$ 8.2 (6–44)			
PCS12 T0	36.5 $\pm$ 8.4 (21.7–50.6)	37.6 $\pm$ 7.7 (22.8–51.8)	37.1 $\pm$ 8.0 (21.7–51.8)	0.045	<0.0001	0.066
PCS12 T1	42.6 $\pm$ 6.9 (30.2–52.6)	50.2 $\pm$ 10.0 (26.7–64.6)	46.4 $\pm$ 9.3 (26.7–64.6)			
MCS12 T0	43.3 $\pm$ 10.1 (26.2–59.1)	46.5 $\pm$ 10.7 (19.4–64.6)	44.9 $\pm$ 10.4 (19.4–64.6)	0.138	<0.0001	0.190
MCS12 T1	47.9 $\pm$ 10.7 (27.0–63.5)	54.3 $\pm$ 10.3 (22.7–65.0)	51.2 $\pm$ 10.8 (22.7–65.0)			
KOOS T0	62.9 $\pm$ 9.2 (45–76)	59.8 $\pm$ 7.1 (45–72)	61.4 $\pm$ 8.2 (45–76)	0.403	<0.0001	<0.001
KOOS T1	66.4 $\pm$ 8.6 (51–76)	74.0 $\pm$ 11.0 (48–89)	70.2 $\pm$ 10.5 (58–89)			
OKS T0	29.4 $\pm$ 4.8 (18–36)	27.4 $\pm$ 6.3 (13–38)	28.4 $\pm$ 5.6 (13–38)	0.278	<0.0001	<0.0001
OKS T1	30.6 $\pm$ 3.1 (24–35)	36.6 $\pm$ 8.2 (17–47)	33.6 $\pm$ 6.8 (17–47)			

Control = control group; treatment = treatment group; *n* = number; NRS = Numerical Rating Scale; ODI = Oswestry Disability Index; PCS12 = SF12 Physical Component Dimension; MCS12 = SF12 Mental Component Dimension); KOOS = Knee Injury and Osteoarthritis Outcome Score; OKS = Oxford Knee Score.

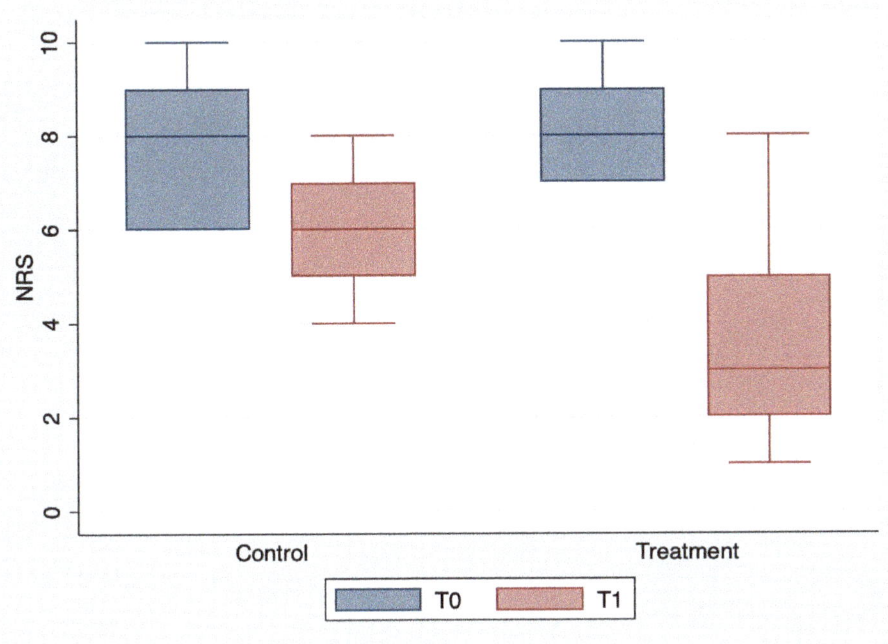

Figure 1. Box plot of NRS values by detection time and group.

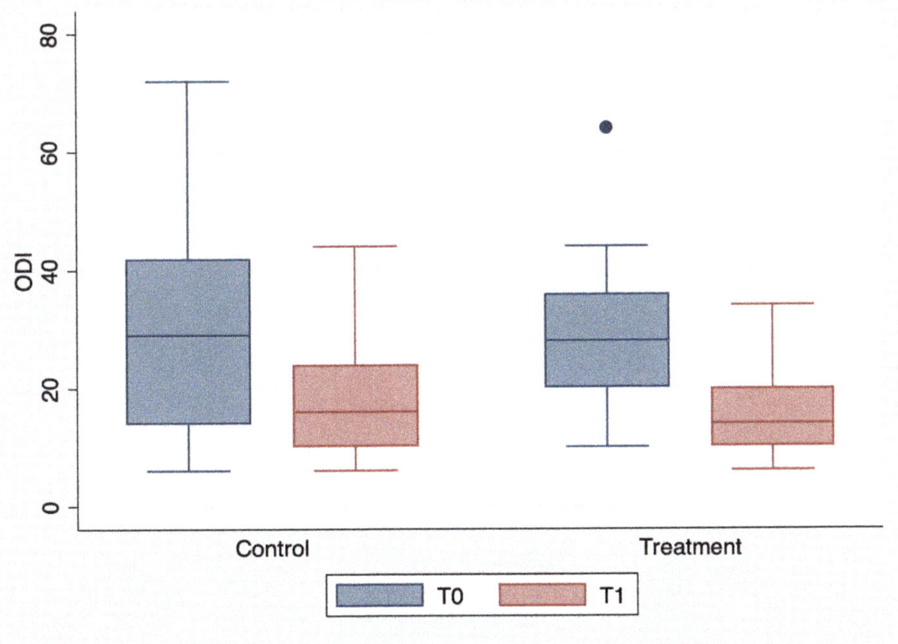

Figure 2. Box plot of ODI values by detection time and group.

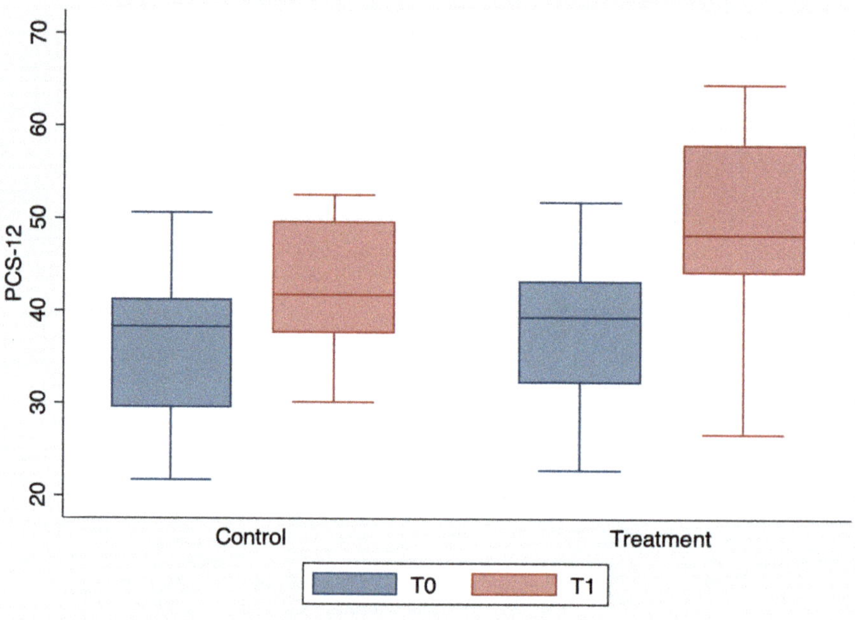

Figure 3. Box plot of PCS-12values by detection time and group.

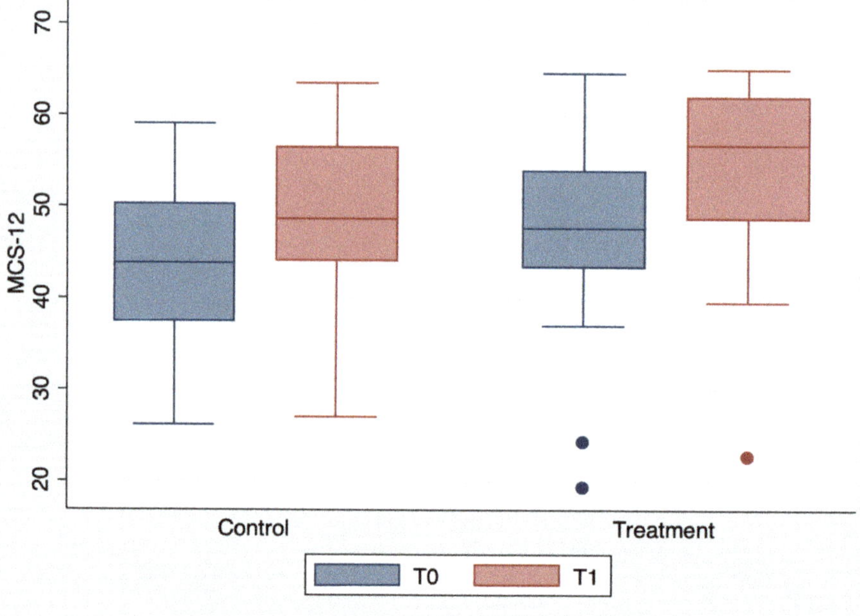

Figure 4. Box plot of MCS-12 values by detection time and group.

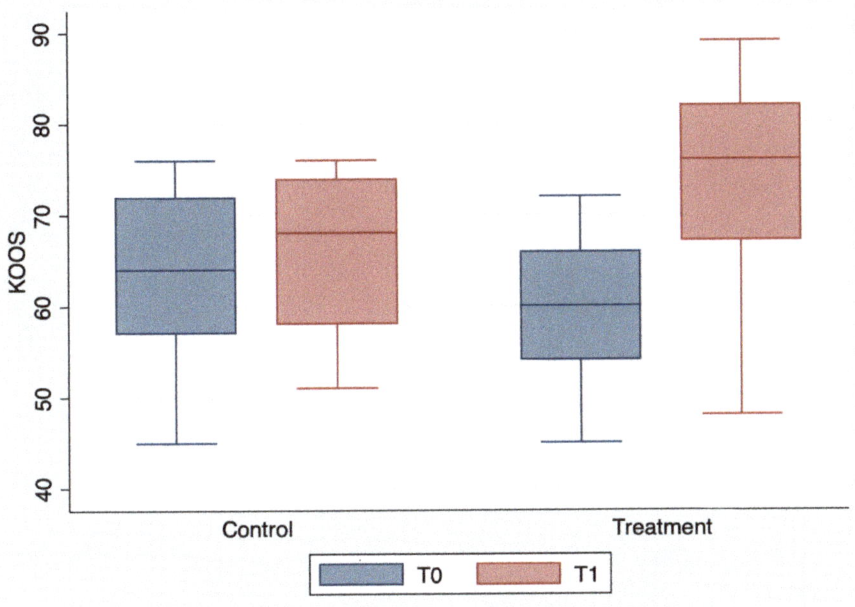

Figure 5. Box plot of KOOS values by detection time and group.

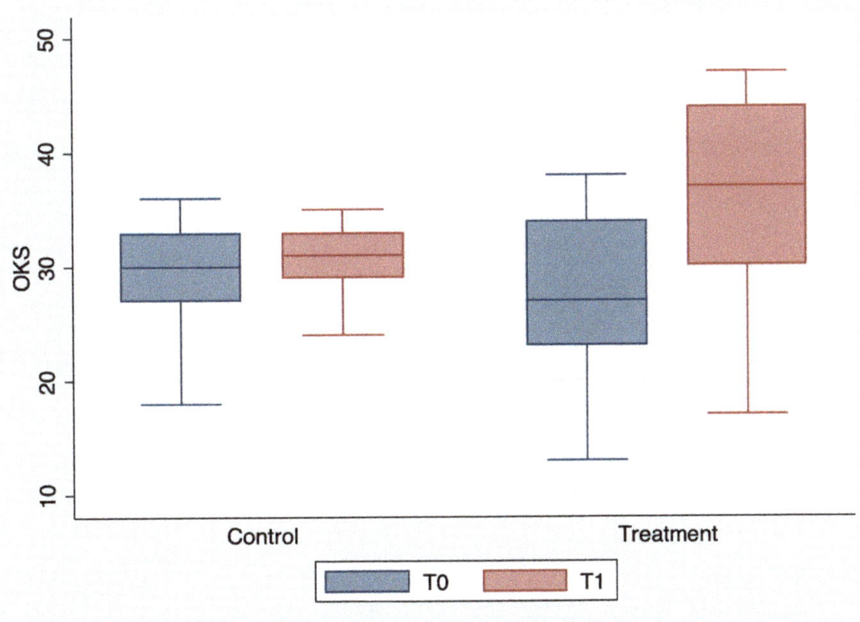

Figure 6. Box plot of OKS values by detection time and group.

Tables 3–8 describe the multivariate linear regression analyses by single outcome. Specifically, in Table 3 a statistically significant improvement in the NRS scores emerged between T0 and T1, attributable solely to the treatment ($p < 0.0001$).

Table 3. Analysis of the determinants of the difference between NRS T1 and NRS T0 in a multivariate linear regression model.

Determinants	Coef.	95%CI	*p*-Value
Group (treatment vs. control)	−2.8	−4.0–1.6	<0.0001
Sex (male vs. female)	0.04	−1.08–1.17	0.938
Age (years)	0.04	−0.06–0.14	0.404
BMI	0.04	−0.07–0.15	0.486

BMI = Body Mass Index; Coef = coefficient; CI = Confidence Interval.

Table 4. Analysis of the determinants of the difference between ODI T1 and ODI T0 in a multivariate linear regression model.

Determinants	Coef.	95%CI	*p*-Value
Group (treatment vs. control)	0.89	−9.0–10.8	0.856
Sex (male vs. female)	−4.70	−13.8–4.4	0.303
Age (years)	−0.19	−0.98–0.61	0.640
BMI	−0.97	−1.88−−0.07	0.036

BMI = Body Mass Index; Coef = coefficient; CI = Confidence Interval.

Table 5. Analysis of the determinants of the difference between PCS-12 T1 and PCS-12 T0 in a multivariate linear regression model.

Determinants	Coef.	95%CI	*p*-Value
Group (treatment vs. control)	8.5	0.6–16.3	0.036
Sex (male vs. female)	0.6	−6.6–7.8	0.864
Age (years)	0.2	−0.4–0.9	0.409
BMI	−0.4	−1.1–0.3	0.036

BMI = Body Mass Index; Coef = coefficient; CI = Confidence Interval.

Table 6. Analysis of the determinants of the difference between MCS-12 T1 and MCS-12 T0 in a multivariate linear regression model.

Determinants	Coef.	95%CI	*p*-Value
Group (treatment vs. control)	3.7	−2.3–9.6	0.218
Sex (male vs. female)	−0.03	−5.50–5.44	0.992
Age (years)	0.10	−0.38–0.58	0.674
BMI	0.08	−0.46–0.62	0.769

BMI = Body Mass Index; Coef = coefficient; CI = Confidence Interval.

Table 7. Analysis of the determinants of the difference between KOOS T1 and KOOS T0 in a multivariate linear regression model.

Determinants	Coef.	95%CI	*p*-Value
Group (treatment vs. control)	11.9	5.8–18.0	<0.0001
Sex (male vs. female)	0.7	−4.9–6.3	0.792
Age (years)	0.2	−0.3–0.6	0.515
BMI	−0.2	−0.8–0.3	0.402

BMI = Body Mass Index; Coef = coefficient; CI = Confidence Interval.

Table 8. Analysis of the determinants of the difference between OKS T1 and OKS T0 in a multivariate linear regression model.

Determinants	Coef.	95%CI	*p*-Value
Group (treatment vs. control)	8.5	5.4–11.7	<0.0001
Sex (male vs. female)	0.4	−2.5–3.3	0.796
Age (years)	0.1	−0.2–0.3	0.511
BMI	−0.1	−0.4–0.2	0.537

BMI = Body Mass Index; Coef = coefficient; CI = Confidence Interval.

In Table 4, a statistically significant improvement in the ODI scores emerged between T0 and T1, attributable exclusively to patients' BMI ($p = 0.036$).

In Table 5, a statistically significant improvement in the PCS-12 scores emerged between T0 and T1, attributable to the treatment ($p < 0.036$) and to BMI ($p < 0.036$).

In Table 6, no statistically significant differences were found between T0 and T1 for MCS-12 values. However, it should be noted that the starting values were already uneven between the two groups, especially with regard to the MCS-12. This is attributable to the fact that this rating scale, especially for the mental dimension, is easily influenced by factors other than knee pain.

In Table 7, a statistically significant improvement in the KOOS scores emerged between T0 and T1, attributable exclusively to the treatment ($p < 0.0001$).

In Table 8, a statistically significant improvement in the OKS scores emerged between T0 and T1, attributable exclusively to the treatment ($p < 0.0001$).

From the analysis of analgesic intake diaries, only a random intake emerged, which settled on an average of 1.0 g/week per group, with a sporadic and not significant distribution among the participants. No side effects were referred.

4. Discussion

The aforementioned results showed that both dietary supplements produced beneficial effects in patients. However, NRS, KOOS and OKS scores had a statistically significant greater improvement in the treatment group, which is the one treated with the dietary supplement containing both hemp seed oil and terpenes. Therefore, this latter seems more effective in relieving KOA pain and improving specific knee function.

As was said previously, hemp seed oil accounts for the increasing scores in both groups [15,16], but in assessing the composition of the two dietary supplements it is likely that the better results of the treatment group, both in terms of pain relief and in terms of joint function, derive from the presence in the one taken by this group of terpenes, more specifically BCP and myrcene. In fact, in 2020, Rao Jiang-Yan et al. showed how, through autophagic activation, BCP is able to alleviate cerebral ischemia/reperfusion injury in mice, highlighting its protective role in animal cells and vessels [28]. Experimental studies showed that BCP reduces pro-inflammatory mediators such as tumor necrosis factor-alfa (TNF-α), interleukin-1β (IL-1β), interleukin-6 (IL-6), and nuclear factor kappa-light-chain-enhancer of activated B cells (NF-κB), thus ameliorating chronic pathologies characterized by inflammation and oxidative stress [29–32]. In 2012, Ou Ming-Chiu et al. recruited 48 women with primary dysmenorrhea with an NRS > 5. On menstrual cycle days, twenty-four patients massaged synthetic fragrances on the abdomen and the other twenty-four patients did the same with BCP-based essential oils. Abdomen massages with BCP-based essential oils provided relief to patients with primary dysmenorrhea, and the duration of menstrual pain was also reduced [33]. Moreover, Shim Ik Hyun et al. showed the effectiveness of BCP in reducing Helicobacter Pylori related gastritis, particularly nausea and epigastric pain [34]. Due to its lipophilicity, BCP is highly lipophilic, so it possesses a good oral bioavailability [35]. Ibrahim et al. demonstrated BCP's ability to act as a significant antinociceptive without any damage to gastric mucosa [36]. In addition, BCP is able to reduce the expression of COX-2 and inducible Nitric Oxide Synthase (iNOS), avoiding NF-κB activation, so analgesia is consequently achieved [37]. The reduction of acute and chronic pain is achieved by BCP because of its interaction with the opioid system [38]. In fact, BCP promotes the release of β-endorphin secondarily affecting the opioid system [36]. Growing evidence highlighted the suitability of BCP for the treatment of chronic inflammation [32], such as the one deriving from osteoarthritis.

Although less documented than BCP, myrcene has also demonstrated its anti-inflammatory properties. In a study conducted by Shamsul et al., this molecule reduced pro-inflammatory cytokines (IL-1β, IL-6, and TNF-α), immunomodulatory factors (interferon gamma (IFNγ), NF-κB and anti-inflammatory markers [interleukin-4 (IL-4), and interleukin-10 (IL-10)] [39]. Interestingly, myrcene is able to act on Transient Receptor Potential Vanilloid 1 (TRPV1), suggesting its potential analgesic action [40]. Similar studies regarding other dietary supplements used to reduce OA pain achieved results in pain control at short follow-ups which were comparable to ours. In particular, a trial that tested the effectiveness of a collagen peptide-based supplement in reducing OA-related lower back pain recorded a 4.1-points reduction according to the Visual Analogical Scale (VAS) in the treatment group after only 3 weeks of intake [16].

Similarly, a recent 8-week randomized double-blind placebo-controlled trial by Wang et al. reached a 2.6-points pain relief according to the Western Ontario and McMaster University (WOMAC) Osteoarthritis index using oral low molecular weight hyaluronic acid in combination with glucosamine and chondroitin on KOA in patients with mild knee pain [41]. Moreover, such evidence is increasingly present in the available literature for supplements apparently less specific for cartilage, but equally valid for the well-known anti-inflammatory power. A systematic review of the nutritional supplement Perna Canaliculus (green-lipped mussel) in the treatment of OA revealed that this molecule could achieve great pain relief at short follow-up according to its ability to counteract joint inflammatory processes [42]. Likewise, cannabidiol demonstrated very encouraging results in counteracting OA-related pain and joint degeneration in both animal models [43] and human studies [44]. Thus, the pain relief we obtained seems to be in line with these results, and was justified by the progressive control of the underlying inflammation.

With regard to the functional outcomes, the ODI and SF-12 scales did not differ appreciably between the two groups, while the KOOS and OKS scales improved significantly in the treatment group. We consider this as the consequence of the fact that ODI and SF-12 are nonspecific for knee evaluation, while KOOS and OKS, as the name suggests (Knee Injury and Osteoarthritis Scale and Oxford Knee Scale), are extremely specific for the pathology we analyzed, which is KOA. Indeed, as is well known, ODI proved to be more suitable for the functional evaluation of spinal disorders [45], while SF-12, which investigates the quality of life, remained less specific and more easily influenced by other factors, especially psychological ones [46]. Furthermore, Tables 4 and 5 suggest that BMI has a relevant impact on these scales, confirming their nonspecific nature.

We assume that the obtained knee functional improvement is due to the fact that a pain-free joint works better. In fact, one of the first goals in the rehabilitation of knee diseases is precisely to intervene on pain in order to improve joint function and range of motion (ROM) as early as possible [47]. Bahr Taylor et al. showed that massages with essential oils (composed of 55 percent of BCP) on the hands of rheumatoid arthritis patients relieved pain, improved finger strength and significantly increased the angle of maximum flexion compared to subjects treated with coconut oil [48]. Moreover, a study conducted by Topp Robert et al. demonstrated how topical treatment with menthol, which is a terpene as well, improved pain and also knee function in patients with KOA compared to the placebo-treated group [49]. In line with our results, a 2020 in vitro and in vivo study demonstrated the anti-inflammatory action of geranol, an acyclic monoterpene which, when taken per os, protects cartilage and improves joint function [50]. Similarly, as early as 2005, the role of ginger extract in suppressing chemokine induction in human synoviocytes was clear [27], and it was then shown to counteract disability and improve functional capacity in adults with OA [51].

Finally, the dietary supplement containing terpenes appears to have useful properties to improve KOA symptoms. It is likely that the beneficial effects of each nutraceutical component are synergistically amplified by combining them. It is important to further investigate the usefulness of nutraceutical supplements as a complementary treatment for OA because they represent a natural alternative to anti-inflammatory drugs with essentially no side effects that can be self-administered by patients [52,53].

The main limitation of this study is the short duration of follow-up. For this reason, further studies are needed to monitor the durability of the benefits noted in the short term over time. Moreover, the outcome measures are self-reported by patients, but it is a mandatory condition when it is necessary to investigate joint function in the activities of daily living. Finally, there was not a placebo group, but this choice was due to the necessity of guaranteeing a treatment for all patients, since all of them suffered from significant knee pain. Nevertheless, future studies will overcome this limitation, in compliance with the necessary ethical rules.

5. Conclusions

In conclusion, the dietary supplement containing terpenes in addition to hemp seed oil is an effective complementary treatment option in patients with KOA for relieving pain and improving joint function. Further studies are needed to prove its efficacy inside multimodal therapies and with longer follow-up periods.

Author Contributions: Conceptualization, G.F. and M.V.R.; methodology, M.M. and M.R.; software, E.C.N., A.D.I., A.S., F.P.B., D.C., S.P. and E.D.C.; validation, S.S. and B.R.; formal analysis, D.M., E.Q., A.G. and V.S.; resources, S.S.; data curation, M.V.R. and E.C.N.; writing—original draft preparation, G.F. and M.V.R.; writing—review and editing, M.R.; supervision, M.M.; project administration, S.S., G.F. and B.R. All authors have read and agreed to the published version of the manuscript.

Funding: This research received no external funding.

Institutional Review Board Statement: The study was conducted in accordance with the Declaration of Helsinki and approved by the Ethics Committee of Albania University, Tiran, Albania (Nr. 587 Prot.–Date: 13 December 2021).

Informed Consent Statement: Informed consent was obtained from all subjects involved in the study.

Data Availability Statement: The datasets used and analyzed during the current study will be made available upon reasonable request to the corresponding author, G.F.

Conflicts of Interest: The authors declare that they have no conflicts of interest.

References

1. Argenson, J.N.; Parratte, S.; Ollivier, M. Approches thérapeutiques de la gonarthrose [Therapeutic approaches to gonarthrosis]. *Rev. Infirm.* **2016**, *223*, 25–27. [CrossRef] [PubMed]
2. Farì, G.; Pignatelli, G.; Macchiarola, D.; Bernetti, A.; Agostini, F.; Fornaro, M.; Mariconda, C.; Megna, M.; Ranieri, M. Cartilage damage: Clinical approach. *Minerva Orthop.* **2022**, *73*, 20–27. [CrossRef]
3. Della Tommasa, S.; Winter, K.; Seeger, J.; Spitzbarth, I.; Brehm, W.; Troillet, A. Evaluation of Villus Synovium From Unaffected Metacarpophalangeal Joints of Adult and Juvenile Horses. *J. Equine Vet. Sci.* **2021**, *102*, 103637. [CrossRef]
4. Kwiatkowski, K. Choroba zwyrodnieniowa stawu kolanowego—epidemiologia i czynniki ryzyka [Gonarthrosis—epidemiology and risk factors]. *Pol. Merkur. Lekarski.* **2004**, *17*, 410–414. [PubMed]
5. Pape, D.; Tischer, T. Kniearthrose des jungen Patienten [Osteoarthritis of the knee in young patients]. *Orthopade* **2021**, *50*, 345. [CrossRef] [PubMed]
6. Jang, S.; Lee, K.; Ju, J.H. Recent Updates of Diagnosis, Pathophysiology, and Treatment on Osteoarthritis of the Knee. *Int. J. Mol. Sci.* **2021**, *22*, 2619. [CrossRef]
7. Michael, J.W.; Schlüter-Brust, K.U.; Eysel, P. The epidemiology, etiology, diagnosis, and treatment of osteoarthritis of the knee. *Dtsch. Arztebl. Int.* **2010**, *107*, 152–162. [CrossRef]
8. Hussain, S.M.; Neilly, D.W.; Baliga, S.; Patil, S.; Meek, R. Knee osteoarthritis: A review of management options. *Scott. Med. J.* **2016**, *61*, 7–16. [CrossRef]
9. Mariconda, C.; Megna, M.; Farì, G.; Bianchi, F.P.; Puntillo, F.; Correggia, C.; Fiore, P. Therapeutic exercise and radiofrequency in the rehabilitation project for hip osteoarthritis pain. *Eur. J. Phys. Rehabil. Med.* **2020**, *56*, 451–458. [CrossRef]
10. Farì, G.; de Sire, A.; Fallea, C.; Albano, M.; Grossi, G.; Bettoni, E.; Di Paolo, S.; Agostini, F.; Bernetti, A.; Puntillo, F.; et al. Efficacy of Radiofrequency as Therapy and Diagnostic Support in the Management of Musculoskeletal Pain: A Systematic Review and Meta-Analysis. *Diagnostics* **2022**, *12*, 600. [CrossRef]

11. Notarnicola, A.; Maccagnano, G.; Farì, G.; Bianchi, F.P.; Moretti, L.; Covelli, I.; Ribatti, P.; Mennuni, C.; Tafuri, S.; Pesce, V.; et al. Extracorporeal shockwave therapy for plantar fasciitis and gastrocnemius muscle: Effectiveness of a combined treatment. *J. Biol. Regul. Homeost. Agents* **2020**, *34*, 285–290. [PubMed]

12. Vicenti, G.; Solarino, G.; Carrozzo, M.; De Giorgi, S.; Moretti, L.; De Crescenzo, A.; Moretti, B. Major concern in the multiligament injuried knee treatment: A systematic review. *Injury* **2019**, *50* (Suppl. 2), S89–S94. [CrossRef] [PubMed]

13. Sachdeva, V.; Roy, A.; Bharadvaja, N. Current Prospects of Nutraceuticals: A Review. *Curr. Pharm. Biotechnol.* **2020**, *21*, 884–896. [CrossRef] [PubMed]

14. Thomas, S.; Browne, H.; Mobasheri, A.; Rayman, M.P. What is the evidence for a role for diet and nutrition in osteoarthritis? *Rheumatology* **2018**, *57* (suppl. 4), iv61–iv74. [CrossRef]

15. Iftikhar, A.; Zafar, U.; Ahmed, W.; Shabbir, M.A.; Sameen, A.; Sahar, A.; Bhat, Z.F.; Kowalczewski, P.Ł.; Jarzębski, M.; Aadil, R.M. Applications of *Cannabis Sativa* L. in Food and Its Therapeutic Potential: From a Prohibited Drug to a Nutritional Supplement. *Molecules* **2021**, *26*, 7699. [CrossRef]

16. Farì, G.; Santagati, D.; Pignatelli, G.; Scacco, V.; Renna, D.; Cascarano, G.; Vendola, F.; Bianchi, F.P.; Fiore, P.; Ranieri, M.; et al. Collagen Peptides, in Association with Vitamin C, Sodium Hyaluronate, Manganese and Copper, as Part of the Rehabilitation Project in the Treatment of Chronic Low Back Pain. *Endocr. Metab. Immune Disord. Drug Targets* **2022**, *22*, 108–115. [CrossRef]

17. Koppel, B.S.; Brust, J.C.; Fife, T.; Bronstein, J.; Youssof, S.; Gronseth, G.; Gloss, D. Systematic review: Efficacy and safety of medical marijuana in selected neurologic disorders: Report of the Guideline Development Subcommittee of the American Academy of Neurology. *Neurology* **2014**, *82*, 1556–1563. [CrossRef]

18. Russo, E.B.; Marcu, J. Cannabis pharmacology: The usual suspects and a few promising leads. *Adv Pharmacol.* **2017**, *80*, 67–134.

19. Hill, K.P. Medical marijuana for treatment of chronic pain and other medical and psychiatric problems: A clinical review. *JAMA* **2015**, *313*, 2474–2483. [CrossRef]

20. Romero-Sandoval, E.A.; Fincham, J.E.; Kolano, A.L.; Sharpe, B.N.; Alvarado-Vazquez, P.A. Cannabis for chronic pain: Challenges and considerations. *Pharmacotherapy* **2018**, *38*, 651–662. [CrossRef]

21. Weston-Green, K. The United Chemicals of Cannabis: Beneficial effects of Cannabis phytochemicals on the brain and cognition. In *Recent Advances in Cannabinoid Research*; Costain, W., Laprairie, R.B., Eds.; IntechOpen: London, UK, 2018; pp. 83–100.

22. Liktor-Busa, E.; Keresztes, A.; LaVigne, J.; Streicher, J.M.; Largent-Milnes, T.M. Analgesic Potential of Terpenes Derived from Cannabis sativa. *Pharmacol. Rev.* **2021**, *73*, 98–126. [CrossRef] [PubMed]

23. Sharma, C.; Al Kaabi, J.M.; Nurulain, S.M.; Goyal, S.N.; Kamal, M.A.; Ojha, S. Polypharmacological properties and therapeutic potential of b-caryophyllene: A dietary phytocannabinoid of pharmaceutical promise. *Curr. Pharm. Des.* **2016**, *22*, 3237–3264. [CrossRef] [PubMed]

24. Gertsch, J.; Leonti, M.; Raduner, S.; Racz, I.; Chen, J.Z.; Xie, X.Q.; Altmann, K.H.; Karsak, M.; Zimmer, A. Beta-caryophyllene is a dietary cannabinoid. *Proc. Natl. Acad. Sci. USA* **2008**, *105*, 9099–9104. [CrossRef]

25. Whiteside, G.; Lee, G.; Valenzano, K. The Role of the Cannabinoid CB2 Receptor in Pain Transmission and Therapeutic Potential of Small Molecule CB2 Receptor Agonists. *Curr. Med. Chem.* **2007**, *14*, 917–936. [CrossRef] [PubMed]

26. Rufino, A.T.; Ribeiro, M.; Sousa, C.; Judas, F.; Salgueiro, L.; Cavaleiro, C.; Mendes, A.F. Evaluation of the anti-inflammatory, anti-catabolic and pro-anabolic effects of E-caryophyllene, myrcene and limonene in a cell model of osteoarthritis. *Eur. J. Pharmacol.* **2015**, *750*, 141–150. [CrossRef] [PubMed]

27. Phan, P.V.; Sohrabi, A.; Polotsky, A.; Hungerford, D.S.; Lindmark, L.; Frondoza, C.G. Ginger extract components suppress induction of chemokine expression in human synoviocytes. *J. Altern. Complement. Med.* **2005**, *11*, 149–154. [CrossRef] [PubMed]

28. Rao, J.Y.; Wang, Q.; Wang, Y.C.; Xiang, F.; Tian, X.C.; Liu, D.H.; Dong, Z. [β-caryophyllene alleviates cerebral ischemia/reperfusion injury in mice by activating autophagy]. *Zhongguo Zhong Yao Za Zhi* **2020**, *45*, 932–936.

29. Mattiuzzo, E.; Faggian, A.; Venerando, R.; Benetti, A.; Belluzzi, E.; Abatangelo, G.; Ruggieri, P.; Brun, P. In Vitro Effects of Low Doses of β-Caryophyllene, Ascorbic Acid and d-Glucosamine on Human Chondrocyte Viability and Inflammation. *Pharmaceuticals* **2021**, *14*, 286. [CrossRef]

30. Irrera, N.; D'Ascola, A.; Pallio, G.; Bitto, A.; Mazzon, E.; Mannino, F.; Squadrito, V.; Arcoraci, V.; Minutoli, L.; Campo, G.M.; et al. β-Caryophyllene Mitigates Collagen Antibody Induced Arthritis (CAIA) in Mice Through a Cross-Talk between CB2 and PPAR-γ Receptors. *Biomolecules* **2019**, *9*, 326. [CrossRef]

31. Bernetti, A.; Agostini, F.; Alviti, F.; Giordan, N.; Martella, F.; Santilli, V.; Paoloni, M.; Mangone, M. New Viscoelastic Hydrogel Hymovis MO.RE. Single Intra-articular Injection for the Treatment of Knee Osteoarthritis in Sportsmen: Safety and Efficacy Study Results. *Front. Pharmacol.* **2021**, *12*, 673988. [CrossRef]

32. Scandiffio, R.; Geddo, F.; Cottone, E.; Querio, G.; Antoniotti, S.; Gallo, M.P.; Maffei, M.E.; Bovolin, P. Protective Effects of (E)-β-Caryophyllene (BCP) in Chronic Inflammation. *Nutrients* **2020**, *12*, 3273. [CrossRef] [PubMed]

33. Ou, M.C.; Hsu, T.F.; Lai, A.C.; Lin, Y.T.; Lin, C.C. Pain relief assessment by aromatic essential oil massage on outpatients with primary dysmenorrhea: A randomized, double-blind clinical trial. *J. Obstet. Gynaecol. Res.* **2012**, *38*, 817–822. [CrossRef] [PubMed]

34. Shim, H.I.; Song, D.J.; Shin, C.M.; Yoon, H.; Park, Y.S.; Kim, N.; Lee, D.H. [Inhibitory Effects of β-caryophyllene on *Helicobacter pylori* Infection: A Randomized Double-blind, Placebo-controlled Study]. *Korean J. Gastroenterol.* **2019**, *74*, 199–204. [CrossRef]

35. Fidyt, K.; Fiedorowicz, A.; Strządała, L.; Szumny, A. B-Caryophyllene and B-Caryophyllene Oxide—Natural compounds of anticancer and analgesic properties. *Cancer Med.* **2016**, *5*, 3007–3017. [CrossRef] [PubMed]

36. Ibrahim, M.M.; Porreca, F.; Lai, J.; Albrecht, P.J.; Rice, F.L.; Khodorova, A.; Davar, G.; Makriyannis, A.; Vanderah, T.W.; Mata, H.P.; et al. CB2 cannabinoid receptor activation produces antinociception by stimulating peripheral release of endogenous opioids. *Proc. Natl. Acad. Sci. USA* **2005**, *102*, 3093–3098. [CrossRef] [PubMed]
37. Fernandes, E.S.; Passos, G.F.; Medeiros, R.; da Cunha, F.M.; Ferreira, J.; Campos, M.M.; Pianowski, L.F.; Calixto, J.B. Anti-inflammatory effects of compounds alpha-humulene and (-)-trans-caryophyllene isolated from the essential oil of Cordia verbenacea. *Eur. J. Pharmacol.* **2007**, *569*, 228–236. [CrossRef] [PubMed]
38. Paula-Freire, L.I.G.; Andersen, M.L.; Gama, V.S.; Molska, G.R.; Carlini, E.L.A. The oral administration of trans-caryophyllene attenuates acute and chronic pain in mice. *Phytomedicine* **2014**, *21*, 356–362. [CrossRef] [PubMed]
39. Islam, A.U.S.; Hellman, B.; Nyberg, F.; Amir, N.; Jayaraj, R.L.; Petroianu, G.; Adem, A. Myrcene Attenuates Renal Inflammation and Oxidative Stress in the Adrenalectomized Rat Model. *Molecules* **2020**, *25*, 4492. [CrossRef]
40. Jansen, C.; Shimoda, L.M.N.; Kawakami, J.K.; Ang, L.; Bacani, A.J.; Baker, J.D.; Badowski, C.; Speck, M.; Stokes, A.J.; Small-Howard, A.L.; et al. Myrcene and terpene regulation of TRPV1. *Channels* **2019**, *13*, 344–366. [CrossRef] [PubMed]
41. Wang, S.J.; Wang, Y.H.; Huang, L.C. The effect of oral low molecular weight liquid hyaluronic acid combination with glucosamine and chondroitin on knee osteoarthritis patients with mild knee pain: An 8-week randomized double-blind placebo-controlled trial. *Medicine* **2021**, *100*, e24252. [CrossRef]
42. Brien, S.; Prescott, P.; Coghlan, B.; Bashir, N.; Lewith, G. Systematic review of the nutritional supplement Perna Canaliculus (green-lipped mussel) in the treatment of osteoarthritis. *QJM* **2008**, *101*, 167–179. [CrossRef] [PubMed]
43. Verrico, C.D.; Wesson, S.; Konduri, V.; Hofferek, C.J.; Vazquez-Perez, J.; Blair, E.; Dunner, K., Jr.; Salimpour, P.; Decker, W.K.; Halpert, M.M. A randomized, double-blind, placebo-controlled study of daily cannabidiol for the treatment of canine osteoarthritis pain. *Pain* **2020**, *161*, 2191–2202. [CrossRef] [PubMed]
44. Vela, J.; Dreyer, L.; Petersen, K.K.; Arendt-Nielsen, L.; Duch, K.S.; Kristensen, S. Cannabidiol treatment in hand osteoarthritis and psoriatic arthritis: A randomized, double-blind, placebo-controlled trial. *Pain* **2022**, *163*, 1206–1214. [CrossRef] [PubMed]
45. Fairbank, J.C.; Pynsent, P.B. The Oswestry Disability Index. *Spine* **2000**, *25*, 2940–2952. [CrossRef]
46. Lera, L.; Márquez, C.; Saguez, R.; Moya, M.O.; Angel, B.; Albala, C. Calidad de vida en personas mayores con depresión y dependencia funcional: Validez del cuestionario SF-12 [Quality of life of older people with depression and dependence: Validity of the SF-12 (short form health survey) questionnaire]. *Rev. Med. Chil.* **2021**, *149*, 1292–1301. [CrossRef]
47. Werner, S. Anterior knee pain: An update of physical therapy. *Knee Surg. Sports Traumatol. Arthrosc.* **2014**, *22*, 2286–2294. [CrossRef]
48. Bahr, T.; Allred, K.; Martinez, D.; Rodriguez, D.; Winterton, P. Effects of a massage-like essential oil application procedure using Copaiba and Deep Blue oils in individuals with hand arthritis. *Complement. Ther. Clin. Pract.* **2018**, *33*, 170–176. [CrossRef]
49. Topp, R.; Brosky, J.A., Jr.; Pieschel, D. The effect of either topical menthol or a placebo on functioning and knee pain among patients with knee OA. *J. Geriatr. Phys. Ther.* **2013**, *36*, 92–99. [CrossRef]
50. Wu, Y.; Wang, Z.; Fu, X.; Lin, Z.; Yu, K. Geraniol-mediated osteoarthritis improvement by down-regulating PI3K/Akt/NF-κB and MAPK signals: In vivo and in vitro studies. *Int. Immunopharmacol.* **2020**, *86*, 106713. [CrossRef] [PubMed]
51. Leach, M.J.; Kumar, S. The clinical effectiveness of Ginger (Zingiber Officinale) in adults with osteoarthritis. *JBI Libr. Syst. Rev.* **2008**, *6*, 310–323. [CrossRef]
52. Tenti, S.; Mondanelli, N.; Bernetti, A.; Mangone, M.; Agostini, F.; Capassoni, M.; Cheleschi, S.; De Chiara, R.; Farì, G.; Frizziero, A.; et al. Impact of Covid-19 pandemic on injection-based practice: Report from an Italian multicenter and multidisciplinary survey. *Ann. Ig.* **2022**, *34*, 501–514. [PubMed]
53. Bernetti, A.; Farì, G.; Mangone, M.; Fiore, P.; Santilli, V.; Paoloni, M.; Agostini, F. Medical management of osteoarthritis during the COVID-19 pandemic: A challenge for the present and the future. *Ann. Di Ig. Med. Prev. E Di Comunita* **2022**, *34*, 184–189.

Article

Utilizing Graphical Analysis of Chest Radiographs for Primary Screening of Osteoporosis

Soichiro Saeki [1,2,*], Kouichi Yamamoto [3], Rie Tomizawa [4,5], Szilvia Meszaros [6], Csaba Horvath [6], Luca Zoldi [7], Helga Szabo [7], Adam Domonkos Tarnoki [4,7], David Laszlo Tarnoki [4,7], Takayuki Ishida [8] and Chika Honda [4,9,*]

[1] Center Hospital of the National Center for Global Health and Medicine, Tokyo 162-8655, Japan
[2] Department of Global and Innovative Medicine, Graduate School of Medicine, Osaka University, Osaka 565-0871, Japan
[3] Department of Radiological Sciences, Faculty of Health Sciences, Morinomiya University of Medical Sciences, Osaka 559-8611, Japan
[4] Center for Twin Research, Graduate School of Medicine, Osaka University, Osaka 565-0871, Japan
[5] School of Nursing, Graduate School of Nursing, Osaka Metropolitan University, Osaka 545-8585, Japan
[6] Department of Internal Medicine and Oncology, Semmelweis University, 1082 Budapest, Hungary
[7] Medical Imaging Centre, Semmelweis University, 1082 Budapest, Hungary
[8] Department of Medical Physics & Engineering, Graduate School of Medicine, Osaka University, Osaka 565-0871, Japan
[9] Department of Public Health Nursing, Shiga University of Medical Science, Otsu 520-2192, Japan
*   Correspondence: sosaeki@hosp.ncgm.go.jp (S.S.); chikah@belle.shiga-med.ac.jp (C.H.); Tel.: +81-3-3202-7181 (S.S.)

Citation: Saeki, S.; Yamamoto, K.; Tomizawa, R.; Meszaros, S.; Horvath, C.; Zoldi, L.; Szabo, H.; Tarnoki, A.D.; Tarnoki, D.L.; Ishida, T.; et al. Utilizing Graphical Analysis of Chest Radiographs for Primary Screening of Osteoporosis. *Medicina* **2022**, *58*, 1765. https://doi.org/10.3390/medicina58121765

Academic Editors: Milica Lazovic and Dejan Nikolić

Received: 20 October 2022
Accepted: 28 November 2022
Published: 30 November 2022

Publisher's Note: MDPI stays neutral with regard to jurisdictional claims in published maps and institutional affiliations.

Abstract: *Background and Objectives*: Osteoporosis is a major risk of fractures, harming patients' quality of life. Dual-energy X-ray absorptiometry (DXA), which can detect osteoporosis early, is too expensive to be conducted on a regular basis. Therefore, we aimed to evaluate a screening method using chest radiographs developed in Japan applied to another population. *Materials and Methods*: Fifty-five patients who had a chest radiograph and DXA and applied within three months of each test were recruited from the patient database of Semmelweis University (Budapest, Hungary). Graphical analysis of the chest radiographs was conducted to identify the ratio of the cortical bone in the clavicle of each patient. Two researchers performed the analysis, and multiple regression was conducted to determine the bone mineral density of each patient provided by DXA. *Results*: The Pearson correlation between two examiners' determinations of the cortical bone ratio was 0.769 ($p < 0.001$). The multiple regression model proved to be statistically significant in identifying osteoporosis, but the model adopted for the Hungarian population was different compared to the Japanese population. *Conclusions*: This simple, economic Japanese graphical analysis method for chest radiographs may be feasible in detecting osteoporosis. Further studies with a larger population of patients with greater variety of ethnicity would be of value in improving the accuracy of this model.

Keywords: bone fractures; bone mineral density; prevention; aging society; Japan; Hungary

1. Introduction

Osteoporosis is associated with decreased bone mass, associated microarchitectural deterioration, and fragility fractures [1]. It is a widespread disease affecting mainly elderly patients, and it is associated with inadequate nutrition, inappropriate calcium and vitamin D intake, irregular menstrual cycles, and lack of physical exercise. Osteoporosis remains a significant public health problem, mainly because it is largely underdiagnosed and undertreated [2]. Osteoporosis is indicated to have a major impact on mortality [3] as well as harming patients' quality of life [4] by making them more "immobile" [5]. In Japan, as the population ages, cases of osteoporosis have been on the rise according to a recent local study [6], and is believed to continue to increase—not only limited to the female population [7]. The prevalence of osteoporosis is estimated to be high in high-income countries [8] other than Japan as well.

The gold standard for diagnosing osteoporosis is the measurement of bone mineral density (BMD) by dual-energy X-ray absorptiometry (DXA). However, DXA requires specific, expensive equipment, as well as well-trained technicians. Furthermore, Medicare payments have cut DXA checkup for osteoporosis [9], leading to low rates of DXA screening in the United States [10]. Hence, the necessity for a new, low-cost, simple screening method of osteoporosis is of great necessity.

Previously, Kumar and Anburajan [11] reported a method of grouping patients into low and high BMD from the clavicle cortical bone length ratio. As this method required a quantitative method to determine the margins between the cortical and cancellous bone, Ishikawa et al. [12] reported a method for determining the BMD of patients from the clavicle by graphical analysis. However, this method has only been validated among patient data in Japan, which utilizes the young adult mean as the key indicator for the diagnosis of osteoporosis [13,14], unlike the other countries using Z-scores and T-scores [14]. Therefore, we conducted a study to evaluate a method of screening chest X-rays obtained from patients in Hungary to comply with global standards.

2. Materials and Methods

2.1. Subjects, Study Design

This is a single-centered, retrospective study conducted in Hungary. The study data were derived from patient records and analyzed for secondary use. From the patient database of Semmelweis University (Budapest, Hungary), 55 patients who had a chest radiograph and DXA and applied within three months of each test were identified and included in this study. Chest radiographs were performed on the following X-ray devices: 7X PRO 100-HF 650 (7x Orvostechnika Ltd., Budapest, Hungary), 7X Super 750B (7x Orvostechnika Ltd., Budapest, Hungary), GE Discovery XR 656 (GE Healthcare, Milwaukee, WI, USA).

Bone mineral density was measured with GE Lunar Prodigy (GE Healthcare, Milwaukee, WI, USA) dual-energy X-ray absorptiometry. As shown in Figure 1, a total of six patients were excluded: three patients as optimal location for the analysis could not be identified; two patients as fundamental data were missing; and one patient as the patient's chest radiograph could not be identified. In addition, of the 49 patients included in the study, four patients were found with data of the left radius missing and were excluded from the analysis of the radius.

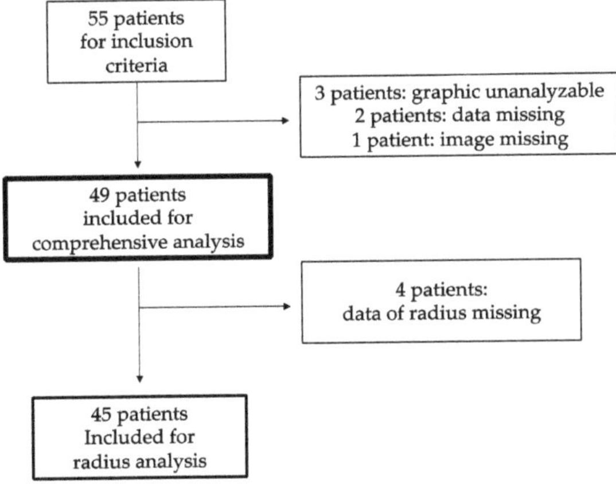

Figure 1. Inclusion and exclusion criteria.

This study protocol was approved by the ethical committee of the Graduate School of Medicine, Osaka University (approval number: 15569-6, approved 17 August 2022) and the Semmelweis University Regional and Institutional Committee of Science and Research Ethics (approval number: 14/2019, approved 15 February 2019).

2.2. Graphical Analysis

Graphical analysis was conducted under the methods developed by Ishikawa et al. [12]

First, the shade of the clavicle was extracted from the chest X-ray. The contrast-limited adaptive histogram equalization conducted by using open source library Fiji v 1.51i [15] created by the National Institute of Health for ImageJ to increase the contrast and to conduct Canny edge detection.

After the filtering, a region of the proximal clavicle where the clavicle became horizontal to the axis of the radiogram was determined by visual judgement, and the edge pixels were extracted by hand. The procedures for drawing perpendicular lines across the upper and lower margins of the clavicle were decided so that the line was between 90 ± 10 degrees from the lines that were drawn to the lower margin of the clavicle from its origin (Figure 2).

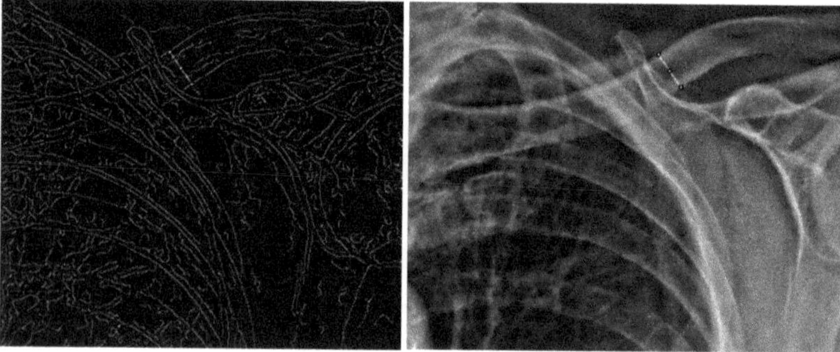

Figure 2. Graphical analysis conducted with Fiji. The region of interest (yellow line) was determined using Canny edge detection, while the pixels were analyzed under enhanced local contrast.

The pixels defined on the perpendicular line from the previous step were used to create a pixel value profile, and an eight-order function was used to fit the approximated curve. The gradients of the tangent lines for the approximated curve were defined in the areas between the upper and lower clavicle margins. The first part where the tangent gradient became the minimum after the first local maximal value was defined as the upper margin of the cortical bone and the cancellous bone, and the distance between that point from the upper clavicle margin was defined as the upper clavicle cortical bone length (CL). The point over the local minimal value with the largest tangent gradient was defined as the lower margin between the cortical bone and the cancellous bone, and the distance between the point of the lower clavicle margin was defined as the lower CL. Clavicle cortical bone-length ratio (CLR) was defined by dividing the upper CL and lower CL by the width of the short axis of the clavicle. From the pixel value profile defined from the previous step, the average CLR was adopted from the three lines used for the analysis.

2.3. Statistical Analysis

We presented each continuous variable with its mean and standard deviation (SD), and each categorical variable is represented by numbers and percentages. A *t*-test was applied to the continuous variables, while a chi-squared test was applied to the categorical variables.

Two independent examiners were assigned to analyze the radiographs. The two examiners conducted their analysis on each side of the clavicle of each patient. Pearson correlation analysis was conducted to validate the results of the CLR between the two examiners. A multiple regression logistics analysis was conducted using the parameters sex, CLR, age [year], body weight [kg], height [cm], BMI (body mass index) [kg/m^2].

EZR (Saitama Medical Center, Jichi Medical University, Saitama, Japan) [16], a graphical user interface for R (version 3.6.1) [17] was used to perform statistical analysis. *p*-value less than 0.05 was determined as statistically significant.

3. Results

Table 1 describes the demographics of the patients included in the study. The average age was 65.3 years, and the average BMI was 26.3. The Pearson correlation between the two examiners' measured CLR was 0.769 (*p* < 0.001).

Table 1. Demographics of patients included in the study.

n = 49 (Male 15)	Average	SD	Min	Max
Age	65.3	13.4	25.0	85.0
BMI	26.3	5.8	17.0	39.4
Height (cm)	161.3	10.4	144.0	188.0
Weight (kg)	68.3	14.7	40.0	105.0
CLR	0.25	0.067	0.15	0.41

Table legends: min: minimum; max: maximum; BMI: body mass index; CLR: clavicle cortical bone length ratio; SD: standard deviation.

Results of the DXA and Z-scores obtained from the DXA are shown in Table 2. The study population showed an average Z-score less than zero in all four regions of the DXA measurement.

Table 2. BMD measured by DXA with Z-scores.

		n	Average	SD	Min	Max
L2-4	(g/cm^2)		1.04	0.24	0.50	1.60
	Z-score		−0.19	1.79	−3.70	3.60
Femoral neck	(g/cm^2)	49	0.80	0.17	0.49	1.29
	Z-score		−0.37	1.23	−3.10	2.50
Total femur	(g/cm^2)		0.84	0.21	0.43	1.44
	Z-score		−0.32	1.44	−4.10	3.20
Radius	(g/cm^2)	45	0.76	0.16	0.49	1.08
	Z-score		−0.48	1.25	−4.30	1.90

Demographics of the BMD and its Z-scores for the patients included in the study. Table legends: BMD: bone mineral density; DXA: dual energy X-ray absorptiometry; SD: standard deviation.

The results of the logistic regression analysis are shown in Table 3 (L2-4), Table 4 (femoral neck), Table 5 (total femur), and Table 6 (radius). The coefficients serve as the variables that estimate the bone mineral density of each patient.

Table 3. Logistic regression of the algorithm obtained from the Hungarian patients on the L2-L4.

Coefficients				
	Estimate	Std. Error	T Value	Pr (> ∣T∣)
(Intercept)	3.128	2.30	1.361	0.18
Age (year)	0.00466	0.0027	1.725	0.09
BMI (kg/m^2)	−0.06062	0.0435	−1.392	0.17
CLR	0.824	0.511	1.614	0.11
Height (cm)	−0.01963	0.00143	−1.374	0.18
Weight (kg)	0.03125	0.0172	1.815	0.07
Sex (male = 1)	0.159	0.0841	1.886	0.06

Other statistical values include residual standard error: 0.1986 at 42 degrees of freedom; multiple R-squared: 0.4045; adjusted R-squared: 0.32; f-statistic: 4.755 on 6 and 42 DF; *p*-value: 0.0008859. Table legends: BMI: body mass index; CLR: clavicle cortical bone length ratio; Std. Error: standard error.

Table 4. Logistic regression of the algorithm obtained from the Hungarian patients on the femoral neck.

Coefficients					
	Estimate	**Std. Error**	**T Value**	**Pr (> I T I)**	
(Intercept)	1.795	1.51	1.190	0.24	
Age (year)	0.000306	0.00177	0.173	0.86	
BMI (kg/m^2)	−0.0113	0.00286	−0.397	0.69	
CLR	0.933	0.335	2.785	0.008	**
Height (cm)	−0.0101	0.00941	−1.070	0.29	
Weight (kg)	0.00930	0.0113	0.824	0.41	
Sex (male = 1)	0.126	0.0552	2.277	0.03	**

Other statistical values include residual standard error: 0.1303 at 42 degrees of freedom; multiple R-squared: 0.4715; adjusted R-squared: 0.40; f-statistic: 6.246 on 6 and 42 DF; *p*-value: 0.00009492. Table legends: BMI: body mass index; CLR: clavicle cortical bone length ratio; Std. Error: standard error; ** $p < 0.001$.

Table 5. Logistic regression of the algorithm obtained from the Hungarian patients on the femoral total Femur.

Coefficients					
	Estimate	**Std. Error**	**T Value**	**Pr (> I T I)**	
(Intercept)	1.416	1.93	0.734	0.47	
Age (year)	0.00112	0.00168	−2.794	0.00803	**
BMI (kg/m^2)	−0.00679	0.0366	−0.186	0.85	
CLR	0.905	0.429	2.109	0.04	*
Height (cm)	−0.00882	0.0120	−0.733	0.4676	
Weight (kg)	0.01000	0.0144	0.692	0.4925	
Sex (male = 1)	0.153	0.0706	2.170	0.0357	*

Other statistical values include residual standard error: 0.1668 at 42 degrees of freedom; multiple R-squared: 0.4575; adjusted R-squared: 0.38; f-statistic: 5.904 on 6 and 42 DF; *p*-value: 0.0001558. Table legends: BMI: body mass index; CLR: clavicle cortical bone length ratio; Std. Error: standard error; * $p < 0.05$; ** $p < 0.001$.

Table 6. Logistic regression of the algorithm obtained from the Hungarian patients on the radius.

Coefficients				
	Estimate	**Std. Error**	**T Value**	**Pr (> I T I)**
(Intercept)	3.128	2.30	1.361	0.18
Age (year)	0.00466	0.0027	1.725	0.09
BMI (kg/m^2)	−0.06062	0.0435	−1.392	0.17
CLR	0.824	0.511	1.614	0.11
Height (cm)	−0.01963	0.00143	−1.374	0.18
Weight (kg)	0.03125	0.0172	1.815	0.07
Sex (male = 1)	0.159	0.0841	1.886	0.06

Other statistical values include residual standard error: 0.1093 on 38 degrees of freedom; multiple R-squared: 0.6099; adjusted R-squared: 0.38; f-statistic: 9.902 on 6 and 38 DF; *p*-value: 0.0000001421. Table legends: BMI: body mass index; CLR: clavicle cortical bone length ratio; Std. Error: standard error.

Our model achieved statistical significance in all regions of the DXA measurement.

4. Discussion

This study aimed to develop a method to utilize chest radiographs for the primary screening of osteoporosis. The results of our study may imply that such methods may be feasible in estimating the BMD of patients who have undergone chest X-rays, which is an important factor for the diagnosis of osteoporosis. To the best of our knowledge, no other study has been conducted to evaluate the status of osteoporosis using graphical imaging using computer analysis.

Osteoporosis is a major public health problem, affecting hundreds of millions of people worldwide. The main clinical consequence of the disease is bone fractures. It is estimated that one in three women [18] and one in five men [19] over the age of fifty worldwide will sustain an osteoporotic fracture. The majority of individuals who have sustained an osteoporosis-related fracture or who are at high risk of fracture are untreated.

Routine chest radiographs obtained for other reasons in various clinical settings can be applied to identify patients at risk of osteoporosis without additional radiation exposure or cost, which could improve osteoporosis screening. For example, in Japan, the screening rate of osteoporosis for women remains low at 4.6% [20], and this simple approach could pave way to identifying potential patients, especially where DXA is not widely available. As DXA requires expensive equipment compared to chest radiographs, this method may be able to promote global health, especially in low- and middle-income countries with a comparative lack of medical resources.

Identifying high-risk groups for osteoporosis using common chest radiographs might increase the disease recognition and prevent osteoporotic fractures. A previous twin study found that BMD is strongly heritable, especially in females in all locations, which highlighted the importance of family history as a risk factor for bone fractures [21]. Public prevention programs could highlight the importance of screening, especially in such risk groups, in preventing fragility fractures. However, population-based screening for osteoporosis is still controversial and has not been implemented [22]. The North American Menopause Society released a position statement on the management of osteoporosis in postmenopausal women in 2021 to reaffirm the importance of screening and assessing risk factors of fractures [23]. Various national societies also have recommendations determining which women should undergo DXA study based on the results of screening tests (questionnaires, fracture risk assessment calculators) [24]. A recent study recommends women be screened for osteoporosis beginning at age 65, while screening for osteoporosis in men should be considered based on the presence of risk factors [25]. The ROSE trial reported that the barriers to population-based screening for osteoporosis appear to be both psychosocial and physical, including factors such as aging, physical impairment, current smoking, and alcohol consumption [22]. Since chest radiographs are routinely used for lung cancer, tuberculosis, and annual workplace suitability screening in some countries among adults, the elderly, and even young populations, we believe that our program could help more efficient screening of those who are at risk of osteoporosis. Although the chest radiography's graphical analysis could not replace DXA for BMD screening, it could be used where DXA has not been performed and chest radiography is readily available.

Our study has two strengths that support the feasibility of the method created by Ishikawa et al. [12]. Firstly, our analysis has been conducted on patients other than the Japanese population. The prevalence of osteoporosis differs from country to country [8], and our manuscript would add to the previous evidence of the Japanese population with another European population. Furthermore, recent evidence has demonstrated that there are health disparities among a variety of diseases [26–28], and this study would also concur with such research. Secondly, we were able to validate the methods with two independent examiners, while the analysis by Ishikawa et al. [12] was conducted by only one personnel member. The strong correlation between the two examiners' results implies that this method may be feasible for different institutions. These two strengths imply that this method may be feasible for different races and ethnicities, and different examinees, which are both important potentials for this method to be applied in clinical situations.

This study has several limitations. Firstly, the nature of the study limits the participants to a relatively osteoporotic population, as the participants were recruited from a selection of hospital patients. This can be seen by the relatively low average of Z-scores (less than zero). Thus, a relatively healthy population may not have been able to participate in our study. However, as our method is thought to be used to screen relatively ill patients, we believe that this aspect of our study may be viewed as a strength, rather than a limitation. Secondly, our model derived from multiple logistic regression could not be verified on a clinical basis, as we were unable to obtain any data that recorded the Z-score of the BMD of the Hungarian population. The results obtained were automatically derived from the measurement machine, and the manufacturer was not able to provide the authors with such data. Thirdly, the sample size of the study remains relatively small, which may have been a reason for reporting models with covariates that were not statistically significant. Fourthly, this study is limited to the Hungarian population. As both genetic and environmental factors are thought to play a role in bone mineral density as well as fractures [21], future studies should include more participants, possibly from a variety of races and ethnicities, from various regions. It may be of interest to conduct studies of race and ethnicity that may not be the majority of the population to identify the influence of environmental factors associated with osteoporosis, although such planning would require substantial effort [29]. Furthermore, as this system relies on a human researcher to conduct the study; this process may be replaceable by artificial intelligence (AI) or machine learning, which could speed up the screening process.

5. Conclusions

In conclusion, our study revealed that a Japanese graphical analysis method using chest radiographs may be feasible in detecting osteoporosis. This method does not require the use of DXA and would be usable in areas under simple analysis. Further studies with a larger population of patients with greater variety of ethnicity would be of value to improving the accuracy of our model.

Author Contributions: Conceptualization, C.H. (Chika Honda) and T.I.; methodology, T.I. and K.Y.; software, S.S.; validation, K.Y. and L.Z.; formal analysis, S.S.; investigation, S.S.; resources, S.M. and C.H. (Csaba Horvath); data curation, S.S.; writing—original draft preparation, S.S.; writing—review and editing, C.H. (Chika Honda), K.Y., R.T., S.M., C.H. (Csaba Horvath), L.Z., H.S., A.D.T., D.L.T. and T.I.; visualization, S.S.; supervision, C.H. (Chika Honda) and T.I.; project administration, C.H. (Chika Honda), A.D.T. and D.L.T.; funding acquisition, C.H. (Chika Honda). All authors have read and agreed to the published version of the manuscript.

Funding: This project was supported by University Grants from the Japanese Ministry of Education, Culture, Sports, Science and Technology; and by JSPS KAKENHI; grant numbers JP19KK0244 and JP17H04134. This research was also supported by Hungarian Academy of Sciences (MTA) Bólyai scholarship, ÚNKP-20-5 and ÚNKP-21-5 New National Excellence Program of the Ministry for Innovation and Technology, from the source of the National Research, Development and Innovation Fund. This work was supported by the Hungarian National Laboratory (under the National Tumorbiology Laboratory project (NLP-17)).

Institutional Review Board Statement: This study was conducted in accordance with the Declaration of Helsinki, and its study protocol was approved by the ethical committee of the Graduate School of Medicine, Osaka University (approval number: 15569-6, approved 17 August 2022) and the Semmelweis University Regional and Institutional Committee of Science and Research Ethics (approval number: 14/2019, approved 15 February 2019).

Informed Consent Statement: Patient consent was waived due to the retrospective nature of this study.

Data Availability Statement: Due to the confidentiality of the patients' personal information, data for this study cannot be shared.

Acknowledgments: We would like to express our gratitude to the patients who participated in the study. Kanako Akada and all technical and secretary staffs at the Center for Twin Research; Department of Global and Innovative Medicine; Osaka University Graduate School of Medicine are acknowledged for their assistance. The authors are thankful to the Erasmus+ (European Commission) and the Osaka University Medical Doctors Student Training Program (MDSTP) for academic and financial support.

Conflicts of Interest: The authors declare no conflict of interest.

References

1. Lane, J.M.; Russell, L.; Khan, S.N. Osteoporosis. *Clin. Orthop. Relat. Res.* **2000**, *372*, 139–150. [CrossRef] [PubMed]
2. Miller, P.D. Management of severe osteoporosis. *Expert Opin. Pharmacother.* **2016**, *17*, 473–488. [CrossRef] [PubMed]

3. Suzuki, T.; Yoshida, H. Low bone mineral density at femoral neck is a predictor of increased mortality in elderly Japanese women. *Osteoporos. Int.* **2010**, *21*, 71–79. [CrossRef] [PubMed]
4. Kumamoto, K.; Nakamura, T.; Suzuki, T.; Gorai, I.; Fujinawa, O.; Ohta, H.; Shiraki, M.; Yoh, K.; Fujiwara, S.; Endo, N.; et al. Validation of the Japanese Osteoporosis Quality of Life Questionnaire. *J. Bone Miner. Metab.* **2010**, *28*, 1. [CrossRef] [PubMed]
5. Shiraki, M.; Kuroda, T.; Shiraki, Y.; Aoki, C.; Sasaki, K.; Tanaka, S. Effects of bone mineral density of the lumbar spine and prevalent vertebral fractures on the risk of immobility. *Osteoporos. Int.* **2010**, *21*, 1545–1551. [CrossRef] [PubMed]
6. Asada, M.; Horii, M.; Ikoma, K.; Goto, T.; Okubo, N.; Kuriyama, N.; Takahashi, K. Hip fractures among the elderly in Kyoto, Japan: A 10-year study. *Arch. Osteoporos.* **2021**, *16*, 30. [CrossRef] [PubMed]
7. Tamaki, J.; Fujimori, K.; Ikehara, S.; Kamiya, K.; Nakatoh, S.; Okimoto, N.; Ogawa, S.; Ishii, S.; Iki, M.; Working Group of Japan Osteoporosis Foundation. Estimates of hip fracture incidence in Japan using the National Health Insurance Claim Database in 2012–2015. *Osteoporos. Int.* **2019**, *30*, 975–983. [CrossRef] [PubMed]
8. Wade, S.W.; Strader, C.; Fitzpatrick, L.A.; Anthony, M.S.; O'Malley, C.D. Estimating prevalence of osteoporosis: Examples from industrialized countries. *Arch. Osteoporos.* **2014**, *9*, 182. [CrossRef]
9. King, A.B.; Fiorentino, D.M. Medicare payment cuts for osteoporosis testing reduced use despite tests' benefit in reducing fractures. *Health Aff.* **2011**, *30*, 2362–2370. [CrossRef]
10. Adams, A.L.; Fischer, H.; Kopperdahl, D.L.; Lee, D.C.; Black, D.M.; Bouxsein, M.L.; Fatemi, S.; Khosla, S.; Orwoll, E.S.; Siris, E.S.; et al. Osteoporosis and Hip Fracture Risk From Routine Computed Tomography Scans: The Fracture, Osteoporosis, and CT Utilization Study (FOCUS). *J. Bone Miner. Res.* **2018**, *33*, 1291–1301. [CrossRef]
11. Kumar, D.A.; Anburajan, M. The role of hip and chest radiographs in osteoporotic evaluation among south Indian women population: A comparative scenario with DXA. *J. Endocrinol. Investig.* **2014**, *37*, 429–440. [CrossRef] [PubMed]
12. Ishikawa, M.; Yamamoto, K.; Matsuzawa, H.; Yamamoto, Y.; Ishida, T. Osteoporosis Screening Using Chest Radiographs. *Med. Imaging Inf. Sci.* **2018**, *35*, 30–34. [CrossRef]
13. Soen, S.; Fukunaga, M.; Sugimoto, T.; Sone, T.; Fujiwara, S.; Endo, N.; Gorai, I.; Shiraki, M.; Hagino, H.; Hosoi, T.; et al. Diagnostic criteria for primary osteoporosis: Year 2012 revision. *J. Bone Miner. Metab.* **2013**, *31*, 247–257. [CrossRef] [PubMed]
14. Kanis, J.A.; Cooper, C.; Rizzoli, R.; Reginster, J.Y.; on behalf of the Scientific Advisory Board of the European Society for Clinical and Economic Aspects of Osteoporosis (ESCEO) and the Committees of Scientific Advisors and National Societies of the International Osteoporosis Foundation (IOF). European guidance for the diagnosis and management of osteoporosis in postmenopausal women. *Osteoporos. Int.* **2019**, *30*, 3–44. [CrossRef] [PubMed]
15. Schindelin, J.; Arganda-Carreras, I.; Frise, E.; Kaynig, V.; Longair, M.; Pietzsch, T.; Preibisch, S.; Rueden, C.; Saalfeld, S.; Schmid, B.; et al. Fiji: An open-source platform for biological-image analysis. *Nat. Methods* **2012**, *9*, 676–682. [CrossRef]
16. Kanda, Y. Investigation of the freely available easy-to-use software 'EZR' for medical statistics. *Bone Marrow Transplant.* **2013**, *48*, 452–458. [CrossRef]
17. R Core Team. *R: A Language and Environment for Statistical Computing*; R Foundation for Statistical Computing: Vienna, Austria, 2019.
18. Melton, L.J., 3rd; Chrischilles, E.A.; Cooper, C.; Lane, A.W.; Riggs, B.L. Perspective. How many women have osteoporosis? *J. Bone Miner. Res.* **1992**, *7*, 1005–1010. [CrossRef]
19. Melton, L.J., 3rd; Atkinson, E.J.; O'Connor, M.K.; O'Fallon, W.M.; Riggs, B.L. Bone density and fracture risk in men. *J. Bone Miner. Res.* **1998**, *13*, 1915–1923. [CrossRef]
20. Orimo, H.; Nakamura, T.; Hosoi, T.; Iki, M.; Uenishi, K.; Endo, N.; Ohta, H.; Shiraki, M.; Sugimoto, T.; Suzuki, T.; et al. Japanese 2011 guidelines for prevention and treatment of osteoporosis—Executive summary. *Arch. Osteoporos.* **2012**, *7*, 3–20. [CrossRef]
21. Piroska, M.; Tarnoki, D.L.; Szabo, H.; Jokkel, Z.; Meszaros, S.; Horvath, C.; Tarnoki, A.D. Strong Genetic Effects on Bone Mineral Density in Multiple Locations with Two Different Techniques: Results from a Cross-Sectional Twin Study. *Medicina* **2021**, *57*, 248. [CrossRef]
22. Rothmann, M.J.; Möller, S.; Holmberg, T.; Højberg, M.; Gram, J.; Bech, M.; Brixen, K.; Hermann, A.P.; Glüer, C.C.; Barkmann, R.; et al. Non-participation in systematic screening for osteoporosis-the ROSE trial. *Osteoporos. Int.* **2017**, *28*, 3389–3399. [CrossRef] [PubMed]
23. Management of osteoporosis in postmenopausal women: The 2021 position statement of The North American Menopause Society. *Menopause* **2021**, *28*, 973–997. [CrossRef] [PubMed]
24. Helmrich, G. Screening for osteoporosis. *Clin. Obstet. Gynecol.* **2013**, *56*, 659–666. [CrossRef]
25. Johnston, C.B.; Dagar, M. Osteoporosis in Older Adults. *Med. Clin. N. Am.* **2020**, *104*, 873–884. [CrossRef] [PubMed]
26. Saeki, S.; Szabo, H.; Tomizawa, R.; Tarnoki, A.D.; Tarnoki, D.L.; Watanabe, Y.; Osaka Twin Research, G.; Honda, C. Lobular Difference in Heritability of Brain Atrophy among Elderly Japanese: A Twin Study. *Medicina* **2022**, *58*, 1250. [CrossRef] [PubMed]
27. Flanagin, A.; Frey, T.; Christiansen, S.L.; AMA Manual of Style Committee. Updated Guidance on the Reporting of Race and Ethnicity in Medical and Science Journals. *JAMA* **2021**, *326*, 621–627. [CrossRef]
28. Saeki, S.; Kusumoto, M. Reporting Race and Ethnicity in Population Health and Clinical Research from Japan. *J. Epidemiol.* **2022**, JE20220244. [CrossRef]
29. Saeki, S. Impact of the "Amendments to the Act of the Protection of Personal Information" to Global Health Research Conducted in Japanese Medical Facilities. *J. Epidemiol.* **2022**, *32*, 438. [CrossRef]

 *medicina*

Article

Reliability, Validity and Temporal Stability of the Serbian Version of the Boston Carpal Tunnel Questionnaire

Darko Bulatovic [1], Dejan Nikolic [2,3,*], Marija Hrkovic [1,2], Tamara Filipovic [1,2], Dragana Cirovic [2,3], Natasa Radosavljevic [4] and Milica Lazovic [4]

[1] Institute for Rehabilitation, 11000 Belgrade, Serbia
[2] Faculty of Medicine, University of Belgrade, 11000 Belgrade, Serbia
[3] Department of Physical Medicine and Rehabilitation, University Children's Hospital, 11000 Belgrade, Serbia
[4] Department of Biomedical Sciences, State University of Novi Pazar, 36300 Novi Pazar, Serbia
* Correspondence: denikol27@gmail.com

Abstract: *Background and Objectives*: The aim of this study was to validate the Serbian version of the Boston Carpal Tunnel Questionnaire (BCTQ) and to evaluate temporal stability for the purpose of its implementation in the evaluation of Serbian patients with carpal tunnel syndrome (CTS). *Materials and Methods*: For the validation of the Serbian version of the BCTQ ($BCTQ_{SR}$), we tested 69 individuals with diagnosed CTS that were referred for a conservative treatment at the Institute for Rehabilitation. Neurophysiological tests were used for the electrophysiological grading (EG) of CTS severity in the study sample. The final version of the $BCTQ_{SR}$ was given to the tested participants from the study on two occasions: test and retest, with a five-day period between the two measurements. *Results*: The mean value for the symptom severity subscale (SSS) of the $BCTQ_{SR}$ was 3.01 ± 0.94; for the functional status subscale (FSS) of the $BCTQ_{SR}$ it was 2.85 ± 1.00. Cronbach's α for the SSS was 0.91 and 0.93 for the FSS. The intraclass correlation coefficients (ICCs) concerning the test–retest were significant ($p < 0.001$) and were 0.949 for the SSS and 0.959 for the FSS. Those with a higher EG grade had higher values of the SSS and FSS but without a statistical significance ($p = 0.103$ and $p = 0.053$, respectively). The intercorrelation of the $BCTQ_{SR}$ subscales (SSS and FSS) on the test was significant ($p < 0.001$) with a correlation coefficient equal to 0.777. *Conclusion*: The Serbian version of the BCTQ ($BCTQ_{SR}$) was successfully culturally adopted. The $BCTQ_{SR}$ was a valid and reliable instrument for the measurement of symptom severity and functional status in adults with CTS. Therefore, it can be used in clinical practice for patients with CTS.

Keywords: carpal tunnel syndrome; Boston carpal tunnel questionnaire; validation; adults

Citation: Bulatovic, D.; Nikolic, D.; Hrkovic, M.; Filipovic, T.; Cirovic, D.; Radosavljevic, N.; Lazovic, M. Reliability, Validity and Temporal Stability of the Serbian Version of the Boston Carpal Tunnel Questionnaire. *Medicina* **2022**, *58*, 1531. https://doi.org/10.3390/medicina58111531

Academic Editor: Lucio Marinelli

Received: 15 September 2022
Accepted: 25 October 2022
Published: 26 October 2022

Publisher's Note: MDPI stays neutral with regard to jurisdictional claims in published maps and institutional affiliations.

1. Introduction

Carpal tunnel syndrome (CTS) is considered to be the most common peripheral nerve entrapment syndrome [1,2]. It is caused by median nerve compression in the wrist region [2]. In the United States (US), the prevalence of the condition is 7.8% [3]; in European population studies, the prevalence ranges between 1 and 7% [4]. In Italy, for those who perform manual work, reports have stated an increase of over 170% in CTS between 2006 and 2010 [3]. Previously, it was stated that CTS is multifactorial with occupational risk factors such as repetitive hand movements, manual forceful exertion, hand–arm transmitted vibration and the bending or twisting of the wrist. Non-occupational risk factors include obesity, thyroid disease, pregnancy, diabetes mellitus, primary amyloidosis and rheumatoid arthritis, which can participate in the development of CTS [5]. Furthermore, it has been noticed that the body mass index (BMI) is an independent risk factor for CTS [6]. It is 3–4 times more likely for women to develop CTS than men and in 50% of cases with CTS both wrists are affected [7]. In the study of Farioli et al., it was stated that several epidemiological studies had been performed in order to evaluate the gender-specific causes of CTS, including hormonal factors, anthropometric parameters, pregnancy and non-occupational biomechanical exposure [8].

Even though CTS is primarily a clinical diagnosis [9], the confirmation is obtained by electrodiagnostic (EDX) studies [10]. Moreover, EDX studies are useful in a severity assessment of CTS and surgery planning [11]; however, the electrodiagnostic CTS severity might not be associated with the clinical severity [12]. The importance of a prompt and adequate diagnosis in patients with CTS is due to the fact that misdiagnosis and delays in establishing a diagnosis can lead to the persistence of symptoms and prolonged functional impairments [13].

Karabinov et al., stated that numerous questionnaires have been developed for the evaluation of upper limb disease, but the Boston Carpal Tunnel Questionnaire (BCTQ) is used most frequently as a disease-specific instrument for CTS [14]. The BCTQ is a patient-based outcome measure of symptom severity and functional status, specifically developed for CTS patients [15]. So far, the BCTQ has been validated in many languages, including Greek, Bulgarian, Dutch, Chinese, Portuguese, Turkish, Korean, Spanish, Finnish, Slovak and Arabic [12,14,16–24]. The BCTQ is a self-administered questionnaire and, as such, might eliminate bias and is sensitive to clinical changes even though it is subjective [19]. Furthermore, in the secondary analysis of Jerosch-Herold et al., it was pointed out that the symptom severity subscale (SSS) and functional status subscale (FSS) of the BCTQ should be evaluated as two separate subscales instead of being summed into a total score [25].

The aim of this study was to validate the Serbian version of the BCTQ and to evaluate its temporal stability for the purpose of the implementation of this questionnaire in the evaluation of Serbian patients with CTS.

2. Materials and Methods

2.1. Study Group

For the validation of the Serbian version of the BCTQ (BCTQ$_{SR}$), we tested 69 individuals with diagnosed CTS that were referred for a conservative treatment at the Institute for Rehabilitation. The diagnosis of CTS was made by a board-certified Physical Medicine and Rehabilitation (PM&R) specialist with experience in CTS diagnostics and treatments. The inclusion criteria were native Serbian language-speaking patients with a first-time CTS diagnosis. The exclusion criteria were: age under 18 years; the presence of diabetes mellitus, rheumatoid arthritis, polyneuropathy, pregnancy, hypothyroidism and cervical radiculopathy; and cognitively challenged patients who were unable to fill in the questionnaire. Further variables analyzed were gender, age, occupation, dominant hand, lateralization of symptoms and electrophysiological grading on the right and left hand. Prior to inclusion in the study, the participants were informed and consent was obtained. The study was approved by the Institutional Review Board (No: 02/942-2, 13 September 2022).

2.2. Electrophysiological Grading

Neurophysiological tests were used for the electrophysiological grading (EG) of CTS severity in the study sample. Grade 0 referred to the absence of neurophysiological abnormalities; Grade 1 or very mild CTS were described as present abnormalities only in two sensitive tests, including a palm/wrist median/ulnar comparison, inching and a ring-finger "double peak"; Grade 2 or a mild degree of CTS were referred to as the presence of orthodromic sensory conduction velocity from the index finger to the wrist below 40 m/s along with a median motor terminal latency from the wrist to the abductor pollicis brevis muscle below 4.5 ms; Grade 3 or a moderately severe type of CTS were described if the motor terminal latency of the median nerve was above 4.5 ms and lower than 6.5 ms with a preserved sensory nerve action potential from the index finger; Grade 4 or severe CTS were noticed if the motor terminal latency of the median nerve was above 4.5 ms and below 6.5 ms as well as an absent sensory nerve action potential; Grade 5 or very severe CTS were referred to for those with a motor terminal latency of the median nerve above 6.5 ms; and

Grade 6 or an extremely severe type of CTS were described if the surface motor potential from the abductor pollicis brevis muscle was below 0.2 mV, peak-to-peak [26].

2.3. Boston Carpal Tunnel Questionnaire

The BCTQ is a self-administered instrument composed of two subscales; one measures the severity of the symptoms and the other measures the functional status [12,17,24]. The symptom severity subscale (SSS) consists of 11 items assessing pain, paresthesia, numbness, weakness, nocturnal symptoms and overall functional status. The functional status subscale (FSS) consists of 8 items that assess the hand function during common daily activities. Every item scores between 1 and 5: SSS 1 is considered to be no symptoms and 5 is the worst symptoms; FSS 1 is considered to be no difficulty and 5 is an inability to perform activities at all. The overall SSS and FSS scores are calculated as the mean of the scores for the 11 and 8 individual items, respectively, where higher final scores point to a worse condition representation of the patient.

2.4. Adaptation Process

For the purpose of the translation and cultural adaptation of the BCTQ to the BCTQ$_{SR}$ we followed the recommendations of the American Association of Orthopedic Surgeons (AAOS) [27]. At the initial stage or the forward translation, we engaged two bilingual translators of different profiles and backgrounds whose first language was Serbian to produce two translated versions (T1 and T2). One translator was aware of the concepts being examined in the translated questionnaire whereas the other was neither aware nor informed. At the second stage or the translation synthesis, a bilingual board-certified PM&R specialist synthesized the T1 and T2 translated versions, along with the two translators who had participated in the forward translation. At this stage, an active discussion took place regarding any potential discrepancies, finally reaching a consensus and producing a common version of the translation: T12. At the third stage or the stage of back translation, two bilingual translators with English as their mother tongue were engaged; they were neither aware nor informed of the explored concepts and produced two back translated versions (BT1 and BT2). At the fourth stage, an expert committee was formed to achieve cross-cultural equivalence. The expert committee was composed of two university professors of PM&R and two active specialists of PM&R with clinical practice of more than 5 years and with expertise in CTS as well as the bilingual translators included in forward and back translation processes to achieve a consensus and produce a pre-final version of the BCTQ$_{SR}$. At the fifth stage, the pre-final version of the BCTQ$_{SR}$ was distributed to 15 participants who had been diagnosed with CTS. All feedback was discussed and solved, producing the final version of the BCTQ$_{SR}$ [27]. The final version of the BCTQ$_{SR}$ was given to the test participants from the study on two occasions: a test and a retest, with a five-day period between the two measurements.

2.5. Statistical Analysis

The results were presented as numbers (N) and percentages (%) for the categorical variables and mean values (MV) with a standard deviation (SD) for the continuous variables. Cronbach's α was used to assess the internal consistency. For the test–retest reliability, we used the intraclass correlation coefficient (ICC). Values of Cronbach's α above 0.70 were considered to be acceptable [12]. Reliability, according to the values of the ICCs, was grouped as >0.90 = high, 0.75–0.90 = good, 0.50–0.75 = moderate and <0.50 = poor [28]. The test–retest reliability was further analyzed by Bland–Altman plots. The Pearson correlation coefficient was used to assess the intercorrelations of the SSS and FSS subscales as well as to correlate the subscales with an age. Differences between the subscales, according to the gender and EG grading, were obtained by an independent sample test. A receiver operating characteristic (ROC) curve was used to assess the ability of the subscales to discriminate between individuals with a low EG and those with a high EG. The performance was analyzed by the area under the curve (AUC). The statistical significance was set at $p < 0.05$.

3. Results

The characteristics of the patients are presented in Table 1. Female patients were predominantly represented (85.51%). An office job (46.38%) was the most frequent in the study sample regarding the occupation type. The right-handed were predominant (95.65%) and the localization of symptoms symmetrically on both sides was present in half of the individuals tested (50.72%). Considering the electrophysiological grading, the most frequent was Grade 2 (45.45%), followed by Grade 3 (43.93%) on the right side. The same applied for the left side, where Grade 2 was present in 37.68% and Grade 3 in 30.44% (Table 1). Three patients with bilateral CTS were excluded when the electrophysiological grading of the right hand was performed due to a surgical treatment for CTS on the right hand.

Table 1. Patient characteristics.

Gender ($n = 69$)	
Male, n (%)	10 (14.49)
Female, n (%)	59 (85.51)
Age (MV $\pm$ SD)	55.67 $\pm$ 10.77
Occupation ($n = 69$)	
Physical job, n (%)	20 (28.98)
Office job, n (%)	32 (46.38)
Unemployed, n (%)	6 (8.70)
Retired, n (%)	11 (15.94)
Dominant Hand ($n = 69$)	
Right, n (%)	66 (95.65)
Left, n (%)	1 (1.45)
Ambidextrous	2 (2.90)
Lateralization of Symptoms ($n = 69$)	
Both sides, symmetrically, n (%)	35 (50.72)
Both sides, more right, n (%)	17 (24.64)
Both sides, more left, n (%)	6 (8.70)
Only right side, n (%)	10 (14.49)
Only left side, n (%)	1 (1.45)
Electrophysiological Grading: Right Hand ($n = 66$)	
Grade 0, n (%)	1 (1.52)
Grade 1, n (%)	1 (1.52)
Grade 2, n (%)	30 (45.45)
Grade 3, n (%)	29 (43.93)
Grade 4, n (%)	2 (3.03)
Grade 5, n (%)	1 (1.52)
Grade 6, n (%)	2 (3.03)
Electrophysiological Grading: Left Hand ($n = 69$)	
Grade 0, n (%)	10 (14.49)
Grade 1, n (%)	9 (13.04)
Grade 2, n (%)	26 (37.68)
Grade 3, n (%)	21 (30.44)
Grade 4, n (%)	2 (2.90)
Grade 5, n (%)	1 (1.45)
Grade 6, n(%)	0

MV—Mean value; SD—Standard deviation.

The mean value for the SSS of the BCTQ$_{SR}$ was 3.01 ± 0.94; for the FSS of the BCTQ$_{SR}$ it was 2.85 ± 1.00. Cronbach's α for the SSS was 0.91; for the FSS, it was 0.93. These represented an acceptable internal consistency. All items for the SSS and FSS were above 0.70 (Table 2).

Table 2. Mean values of the BCTQ$_{SR}$ and Cronbach's α values.

BCTQ$_{SR}$ Items	MV $\pm$ SD	Cronbach's α if Item Deleted	Total Cronbach's α
SSS			
1	3.01 ± 1.37	0.89	-
2	2.94 ± 1.43	0.90	-
3	2.74 ± 1.18	0.90	-
4	3.07 ± 1.36	0.90	-
5	2.83 ± 1.39	0.91	-
6	3.28 ± 1.23	0.91	-
7	2.90 ± 1.20	0.90	-
8	3.13 ± 1.25	0.91	-
9	3.45 ± 1.27	0.90	-
10	3.17 ± 1.33	0.90	-
11	2.59 ± 1.28	0.90	-
Total	3.01 ± 0.94	-	0.91
FSS			
1	2.43 ± 1.24	0.93	-
2	2.61 ± 1.20	0.91	-
3	2.81 ± 1.29	0.92	-
4	2.86 ± 1.25	0.92	-
5	3.22 ± 1.29	0.92	-
6	3.19 ± 1.09	0.91	-
7	3.28 ± 1.22	0.92	-
8	2.39 ± 1.20	0.92	-
Total	2.85 ± 1.00	-	0.93

BCTQ$_{SR}$—Boston Carpal Tunnel Questionnaire Serbian Version; MV—Mean value; SD—Standard deviation; SSS—Symptom severity subscale; FSS—Functional status subscale.

The ICCs concerning the test–retest were significant ($p < 0.001$); these were 0.949 for the SSS and 0.959 for the FSS.

The scatterplot graphs are presented in Figure 1. There was a high correlation between the test and the retest using Pearson's correlation both for the SSS (0.951) and the FSS (0.939) (Figure 1).

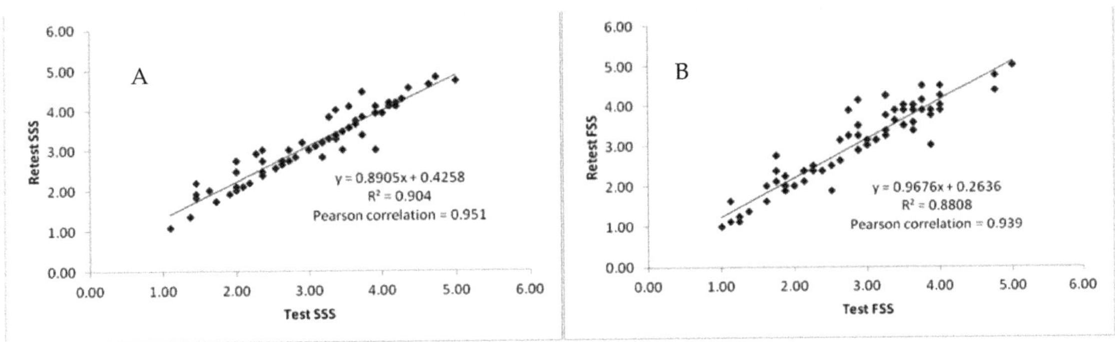

Figure 1. Scatterplot of the test and retest of the symptom severity subscale (SSS) and functional status subscale (FSS) of the Boston Carpal Tunnel Questionnaire Serbian Version (BCTQ$_{SR}$). (**A**): Scatterplot and Pearson's correlation of the test and retest of the SSS. (**B**): Scatterplot and Pearson's correlation of the test and retest of the FSS.

The Bland–Altman plots are presented in Figure 2. The limits of agreement (LoA) for the SSS varied from -0.47 (with 95% CI from -0.36 to -0.58) to 0.67 (with 95% CI from 0.51 to 0.83). For the FSS, the total score varied from -0.53 (with 95% CI from -0.41 to -0.65) to 0.87 (with 95% CI from 0.67 to 1.07) for the time interval between the test and retest, suggesting an acceptable agreement between these two measurements. The average difference for the SSS was 0.10 (with 95% CI from 0.08 to 0.12) and 0.17 for the FSS (from 95% CI from 0.13 to 0.21).

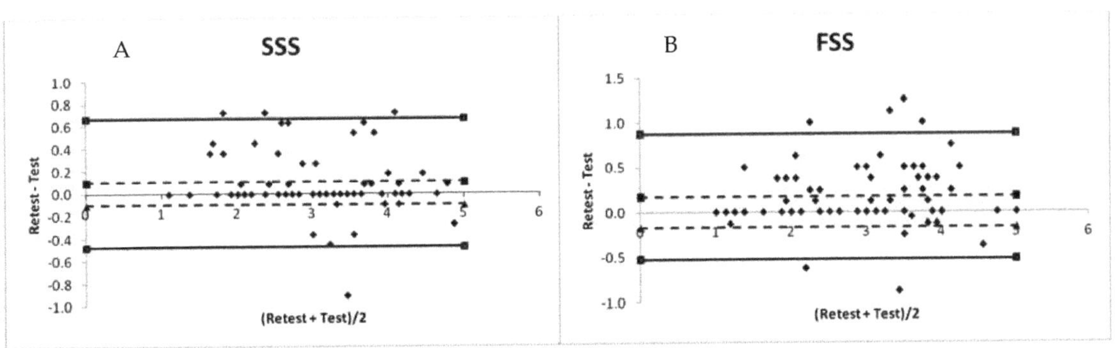

Figure 2. Bland–Altman plots of the test and retest of the symptom severity subscale (SSS) and functional status subscale (FSS) of the Boston Carpal Tunnel Questionnaire Serbian Version BCTQ$_{SR}$. (**A**): Bland–Altman plot of the test and retest of the SSS. (**B**): Bland–Altman plot of the test and retest of the FSS. - - - - negative and positive average difference; - negative and positive 95% confidence interval.

The SSS and FSS scores were higher in females but without a statistical significance ($p = 0.643$ and $p = 0.741$, respectively) on the test. Those with a higher EG grade had higher values of the SSS and FSS but without a statistical significance ($p = 0.103$ and $p = 0.053$, respectively) on the test. Furthermore, there were non-significant correlations between gender and the SSS ($p = 0.719$) and FSS ($p = 0.284$) on the test (Table 3).

Table 3. $BCTQ_{SR}$ association of subscales (SSS and FSS) with gender, age and EG grading on the test.

Tested Variables	SSS		FSS	
		p-Value		*p*-Value
	Gender			
Male (MV ± SD)	2.88 ± 0.99	0.643 *	2.75 ± 1.18	0.741 *
Female (MV ± SD)	3.03 ± 0.94		2.86 ± 0.98	
	Age			
r **	−0.044	0.719 **	0.131	0.284 **
	EG			
1–2 (MV ± SD)	2.80 ± 0.91	0.103 *	2.58 ± 0.97	0.053 *
≥3 (MV ± SD)	3.17 ± 0.94		3.05 ± 0.99	

MV—Mean value; SD—Standard deviation; SSS—Symptom severity subscale; FSS—Functional status subscale; EG—Electrophysiological grading; * Independent sample test; ** Pearson correlation coefficient.

A ROC curve analysis demonstrated that for the SSS, the cut-off value was 3.32, with a sensitivity of 53.8%, a specificity of 70% and an AUC of 0.603 ($p = 0.146$). For the FSS, the cut-off value was 3.06, with a sensitivity of 56.4%, a specificity of 70% and an AUC of 0.644 ($p = 0.042$) (Figure 3).

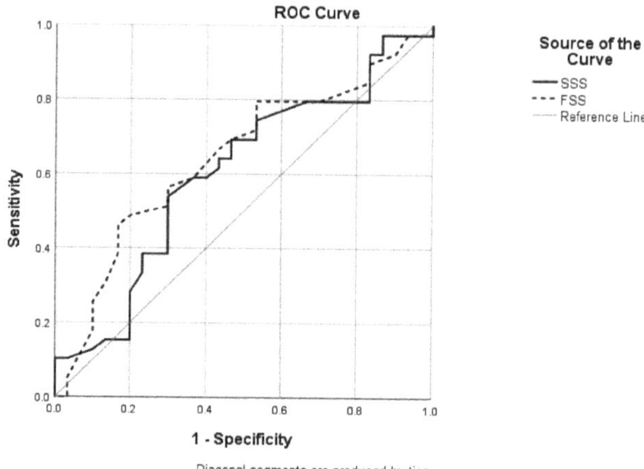

Figure 3. Receiver operating characteristic (ROC) curve for the prediction of a high electrophysiological grading (EG) for symptom severity subscale (SSS) and functional status subscale (FSS) of Boston Carpal Tunnel Questionnaire Serbian Version ($BCTQ_{SR}$) on the test.

The intercorrelation of the $BCTQ_{SR}$ subscales (SSS and FSS) on the test was significant ($p < 0.001$), with a correlation coefficient equal to 0.777.

There was a significant correlation between the SSS items on the test, except for the correlations between SSS Item 6 and SSS Item 2 ($r = 0.192$; $p = 0.114$), SSS Item 6 and SSS Item 4 ($r = 0.189$; $p = 0.121$), SSS Item 6 and SSS Item 5 ($r = 0.062$; $p = 0.610$), SSS Item 8 and SSS Item 5 ($r = 0.216$; $p = 0.074$) and SSS Item 10 and SSS Item 5 ($r = 0.184$; $p = 0.131$). The highest correlation was between Item 5 and Item 4 ($r = 0.858$; $p < 0.001$) and the lowest correlation was between Item 6 and Item 5 ($r = 0.062$; $p = 0.610$) (Table 4).

Table 4. Correlations between SSS items on the test.

BCTQ$_{SR}$ SSS		Item 1	Item 2	Item 3	Item 4	Item 5	Item 6	Item 7	Item 8	Item 9	Item 10	Item 11
Item 1	r	1										
	p											
Item 2	r	0.811	1									
	p	<0.001										
Item 3	r	0.639	0.606	1								
	p	<0.001	<0.001									
Item 4	r	0.630	0.483	0.731	1							
	p	<0.001	<0.001	<0.001								
Item 5	r	0.534	0.385	0.605	0.858	1						
	p	<0.001	0.001	<0.001	<0.001							
Item 6	r	0.346	0.192	0.392	0.189	0.062	1					
	p	0.004	0.114	0.001	0.121	0.610						
Item 7	r	0.645	0.543	0.684	0.533	0.428	0.484	1				
	p	<0.001	<0.001	<0.001	<0.001	<0.001	<0.001					
Item 8	r	0.465	0.333	0.382	0.314	0.216	0.549	0.440	1			
	p	<0.001	0.005	0.001	0.009	0.074	<0.001	<0.001				
Item 9	r	0.659	0.598	0.374	0.304	0.245	0.493	0.455	0.558	1		
	p	<0.001	<0.001	0.002	0.011	0.042	<0.001	<0.001	<0.001			
Item 10	r	0.598	0.747	0.432	0.261	0.184	0.347	0.389	0.439	0.731	1	
	p	<0.001	<0.001	<0.001	0.030	0.131	0.003	0.001	<0.001	<0.001		
Item 11	r	0.493	0.461	0.611	0.440	0.456	0.408	0.730	0.450	0.397	0.398	1
	p	<0.001	<0.001	<0.001	<0.001	<0.001	0.001	<0.001	<0.001	0.001	0.001	

BCTQ$_{SR}$—Boston Carpal Tunnel Questionnaire Serbian Version; SSS—Symptom severity subscale. r: Pearson's correlation.

There was a significant correlation between all FSS items on the test ($p < 0.001$), with the highest correlation between Item 7 and Item 6 (r = 0.833; $p < 0.001$) and the lowest between Item 7 and Item 1 (r = 0.443; $p < 0.001$) (Table 5).

Table 5. Correlations between FSS items on the test.

BCTQ$_{SR}$ FSS		Item 1	Item 2	Item 3	Item 4	Item 5	Item 6	Item 7	Item 8
Item 1	r	1							
	p								
Item 2	r	0.666	1						
	p	<0.001							
Item 3	r	0.512	0.731	1					
	p	<0.001	<0.001						
Item 4	r	0.457	0.596	0.777	1				
	p	<0.001	<0.001	<0.001					
Item 5	r	0.462	0.556	0.591	0.592	1			
	p	<0.001	<0.001	<0.001	<0.001				
Item 6	r	0.471	0.708	0.687	0.646	0.702	1		
	p	<0.001	<0.001	<0.001	<0.001	<0.001			
Item 7	r	0.443	0.654	0.631	0.622	0.547	0.833	1	
	p	<0.001	<0.001	<0.001	<0.001	<0.001	<0.001		
Item 8	r	0.681	0.798	0.609	0.507	0.597	0.662	0.675	1
	p	<0.001	<0.001	<0.001	<0.001	<0.001	<0.001	<0.001	

BCTQ$_{SR}$—Boston Carpal Tunnel Questionnaire Serbian Version; FSS—Functional status subscale. r: Pearson's correlation.

4. Discussion

The translated version of the BCTQ$_{SR}$ was successful because only minor cultural adaptions were needed. The BCTQ$_{SR}$ demonstrated a satisfactory internal consistency and test–retest reliability, with an acceptable agreement between the test and retest for both the SSS and FSS subscales. Furthermore, there was a significant intercorrelation between the BCTQ$_{SR}$ SSS and FSS subscales on the test session.

Regarding the cultural adaptation of the BCTQ$_{SR}$ SSS and FSS items, we also considered the observations from the study of Mendoza-Pulido and Ortiz-Corredor, where weaknesses such as fatigue, sleepiness, unsteadiness or loss of muscle strength could be widely interpreted in patients [29].

The importance of a satisfactory internal consistency refers to the fact that a higher internal consistency is associated with a greater precision or a lower error variance [19]. The results of our study regarding the internal consistencies for the BCTQ$_{SR}$ SSS (Cronbach's $\alpha = 0.91$) and for the BCTQ$_{SR}$ FSS (Cronbach's $\alpha = 0.93$) were in line with previous reports. For example, in the Spanish BCTQ validation for the SSS, the Cronbach's α was 0.909 and for the FSS, the Cronbach's α was 0.872 [21]. In the Greek BCTQ validation, the Cronbach's α was 0.89 for the SSS and the Cronbach's α was 0.93 for the FSS [12]. In the Dutch validation, it was somewhat lower for the SSS (Cronbach's $\alpha = 0.847$) and for the FSS (Cronbach's $\alpha = 0.825$) [16].

In our study, we had no loss of participants between the test and retest of the BCTQ$_{SR}$. Furthermore, no incomplete questionnaires were returned both for the SSS and the FSS on the test and retest. In the study of Leite et al., it was stated that the BCTQ was shown to have good levels of acceptability, with response rates of 90% and above [15].

Even though we found no significant differences in the SSS and FSS scores of the BCTQ$_{SR}$ between genders, females with CTS had higher scores on both BCTQ$_{SR}$ subscales. Our findings differed somewhat when compared with previously reported results, where females with CTS had significantly higher values of both the SSS and FSS [12]. Despite the possibility that females might have a higher sensitivity in the reporting of CTS symptoms and that men might possibly have a higher tolerance for the symptoms [30], the possible explanation for our findings might be in the different cultural and social environment. Furthermore, gender was shown not to significantly correlate with the SSS and FSS of the BCTQ$_{SR}$.

The BCTQ$_{SR}$ scores in our study were higher for patients with CTS who were graded three and above on the EG versus those who were graded from zero to two, but this was without a statistical significance. In the Greek validation, the authors demonstrated significantly increased values for those with Grade 3 and above on the EG when compared with those who had Grades 1 and 2 [12]. The possible explanation for the absence of a significant difference between the tested groups of patients in our study might be due to different perceptions of symptom severity as well as different perceptions of functional changes in the tested patients with CTS.

In the Rash analysis of Multanen et al., regarding the structural validity of the BCTQ, it was noticed that the BCTQ SSS demonstrated multidimensionality whereas the FSS showed a unidimensional structure [22]. Furthermore, these authors pointed to the fact that the question in the BCTQ SSS "How long on average does an episode of pain last during the daytime" demonstrated a non-uniform differential item functioning that favored age whereas Item 7 was shown to favor gender [22].

When considering the correlation between the different items in the BCTQ$_{SR}$ SSS on the test session and the possible explanation for the absence of statistical significances between certain items of the SSS, we referred to the fact that, according to De Kleermaeker et al., the FSS could be considered to be a unidimensional scale whereas the SSS subscale measures three different factors (daytime symptoms, night-time symptoms and operational capacity) [16]. Assuming this, the absence of significant correlations between Item 6 and Item 4, between Item 6 and Item 5 and between Item 8 and Item 5 as well as between Item 10 and Item 5 for the SSS could be explained by the possibility that they belonged to different factors such as "daytime symptoms" for Items 4 and 5 and "night-time symptoms and numbness/tingling" for Item 6, Item 8 and Item 10, as stated in the study of Atroshi et al. [31].

5. Conclusions

The Serbian version of the BCTQ (BCTQ$_{SR}$) was successfully culturally adopted. The BCTQ$_{SR}$ was a valid and reliable instrument for the measurement of symptom severity and functional status in adults with CTS. Thus, it can be used in clinical practice for patients with CTS.

Author Contributions: D.B., D.N. and M.L.: conceptualization, methodology, supervision and writing—original draft; M.H., D.C., N.R. and T.F.: data curation, formal analysis and writing—original draft. All authors have read and agreed to the published version of the manuscript.

Funding: The authors received no financial support for the research and/or authorship.

Institutional Review Board Statement: The study was approved by the Institutional Review Board (No: 02/942-2, 13 September 2022).

Informed Consent Statement: Written informed consent was obtained from each patient.

Data Availability Statement: Original data are available on request from the first author (D.B.)

Conflicts of Interest: The authors declare no conflict of interest concerning the authorship and/or publication of this article.

References

1. Wright, A.R.; Atkinson, R.E. Carpal Tunnel Syndrome: An Update for the Primary Care Physician. *Hawai'i J. Health Soc. Welf.* **2019**, *78* (Suppl. S2), 6–10.
2. Padua, L.; Coraci, D.; Erra, C.; Pazzaglia, C.; Paolasso, I.; Loreti, C.; Caliandro, P.; Hobson-Webb, L.D. Carpal tunnel syndrome: Clinical features, diagnosis, and management. *Lancet Neurol.* **2016**, *15*, 1273–1284. [CrossRef]
3. Scalise, V.; Brindisino, F.; Pellicciari, L.; Minnucci, S.; Bonetti, F. Carpal Tunnel Syndrome: A National Survey to Monitor Knowledge and Operating Methods. *Int. J. Environ. Res. Public Health* **2021**, *18*, 1995. [CrossRef] [PubMed]
4. Middleton, S.D.; Anakwe, R.E. Carpal tunnel syndrome. *BMJ* **2014**, *349*, g6437. [CrossRef] [PubMed]
5. Al Shahrani, E.; Al Shahrani, A.; Al-Maflehi, N. Personal factors associated with carpal tunnel syndrome (CTS): A case-control study. *BMC Musculoskelet. Disord.* **2021**, *22*, 1050. [CrossRef]
6. Lampainen, K.; Shiri, R.; Auvinen, J.; Karppinen, J.; Ryhänen, J.; Hulkkonen, S. Weight-Related and Personal Risk Factors of Carpal Tunnel Syndrome in the Northern Finland Birth Cohort 1966. *J. Clin. Med.* **2022**, *11*, 1510. [CrossRef]
7. Scanlon, A.; Maffei, J. Carpal tunnel syndrome. *J. Neurosci. Nurs.* **2009**, *41*, 140–147. [CrossRef]
8. Farioli, A.; Curti, S.; Bonfiglioli, R.; Baldasseroni, A.; Spatari, G.; Mattioli, S.; Violante, F.S. Observed Differences between Males and Females in Surgically Treated Carpal Tunnel Syndrome Among Non-manual Workers: A Sensitivity Analysis of Findings from a Large Population Study. *Ann. Work Expo. Health* **2018**, *62*, 505–515. [CrossRef]
9. Calandruccio, J.H.; Thompson, N.B. Carpal Tunnel Syndrome: Making Evidence-Based Treatment Decisions. *Orthop. Clin. N. Am.* **2018**, *49*, 223–229. [CrossRef]
10. Alanazy, M.H. Clinical and electrophysiological evaluation of carpal tunnel syndrome: Approach and pitfalls. *Neurosciences* **2017**, *22*, 169–180. [CrossRef]
11. Wipperman, J.; Goerl, K. Carpal tunnel syndrome: Diagnosis and management. *Am. Fam. Physician* **2016**, *94*, 993–999. [PubMed]
12. Bougea, A.; Zambelis, T.; Voskou, P.; Katsika, P.Z.; Tzavara, C.; Kokotis, P.; Karandreas, N. Reliability and Validation of the Greek Version of the Boston Carpal Tunnel Questionnaire. *HAND* **2018**, *13*, 593–599. [CrossRef] [PubMed]
13. Shetty, K.D.; Robbins, M.; Aragaki, D.; Basu, A.; Conlon, C.; Dworsky, M.; Benner, D.; Seelam, R.; Nuckols, T.K. The quality of electrodiagnostic tests for carpal tunnel syndrome: Implications for surgery, outcomes, and expenditures. *Muscle Nerve* **2020**, *62*, 60–69. [CrossRef] [PubMed]
14. Karabinov, V.; Slavchev, S.A.; Georgiev, G.P. Translation and Validation of the Bulgarian Version of the Boston Carpal Tunnel Questionnaire. *Cureus* **2020**, *12*, e10901. [CrossRef]
15. Leite, J.C.D.C.; Jerosch-Herold, C.; Song, F. A systematic review of the psychometric properties of the Boston Carpal Tunnel Questionnaire. *BMC Musculoskelet. Disord.* **2006**, *7*, 78–79. [CrossRef]
16. De Kleermaeker, F.G.C.M.; Levels, M.; Verhagen, W.I.M.; Meulstee, J. Validation of the Dutch Version of the Boston Carpal Tunnel Questionnaire. *Front. Neurol.* **2019**, *10*, 1154. [CrossRef]
17. Lue, Y.-J.; Lu, Y.-M.; Lin, G.-T.; Liu, Y.-F. Validation of the Chinese Version of the Boston Carpal Tunnel Questionnaire. *J. Occup. Rehabil.* **2014**, *24*, 139–145. [CrossRef]

18. de Campos, C.C.; Manzano, G.M.; de Andrade, L.B.; Castelo Filho, A.; Nóbrega, J.A. Translation and validation of an instrument for evaluation of severity of symptoms and the functional status in carpal tunnel syndrome. *Arq. Neuro-Psiquiatr.* **2003**, *61*, 51–55.

19. Sezgin, M.; Incel, N.A.; Sevim, S.; Çamdeviren, H.; As, I.; ErdoĞan, C. Assessment of symptom severity and functional status in patients with carpal tunnel syndrome: Reliability and validity of the Turkish version of the Boston questionnaire. *Disabil. Rehabil.* **2006**, *28*, 1281–1286. [CrossRef]

20. Park, D.-J.; Kang, J.-H.; Lee, J.-W.; Lee, K.-E.; Wen, L.; Kim, T.-J.; Park, Y.-W.; Nam, T.-S.; Kim, M.-S.; Lee, S.-S. Cross-Cultural Adaptation of the Korean Version of the Boston Carpal Tunnel Questionnaire: Its Clinical Evaluation in Patients with Carpal Tunnel Syndrome Following Local Corticosteroid Injection. *J. Korean Med. Sci.* **2013**, *28*, 1095–1099. [CrossRef]

21. Oteo-Álvaro, Á.; Marín, M.T.; Matas, J.A.; Vaquero, J. Validación al castellano de la escala Boston Carpal Tunnel Questionnaire. *Med. Clin.* **2016**, *146*, 247–253. [CrossRef] [PubMed]

22. Multanen, J.; Ylinen, J.; Karjalainen, T.; Kautiainen, H.; Repo, J.; Häkkinen, A. Reliability and Validity of The Finnish Version of The Boston Carpal Tunnel Questionnaire among Surgically Treated Carpal Tunnel Syndrome Patients. *Scand. J. Surg.* **2020**, *109*, 343–350. [CrossRef] [PubMed]

23. Ulbrichtová, R.; Jakušová, V.; Švihrová, V.; Dvorštiaková, B.; Hudečková, H. Validation of the Slovakian version of Boston Carpal Tunnel Syndrome Questionnaire (BCTSQ). *Acta Med.* **2019**, *62*, 105–108. [CrossRef] [PubMed]

24. Hamzeh, H.H.; Alworikat, N.A. Cross cultural adaptation, reliability and construct validity of the Boston Carpal Tunnel Questionnaire in standard Arabic language. *Disabil. Rehabil.* **2021**, *43*, 430–435. [CrossRef] [PubMed]

25. Jerosch-Herold, C.J.; Bland, J.D.P.; Horton, M. Is it time to revisit the Boston Carpal Tunnel Questionnaire? New insights from a Rasch model analysis. *Muscle Nerve* **2021**, *63*, 484–489. [CrossRef]

26. Bland, J.D. A neurophysiological grading scale for carpal tunnel syndrome. *Muscle Nerve* **2000**, *23*, 1280–1283. [CrossRef]

27. Beaton, D.E.; Bombardier, C.; Guillemin, F.; Ferraz, M.B. Guidelines for the Process of Cross-Cultural Adaptation of Self-Report Measures. *Spine* **2000**, *25*, 3186–3191. [CrossRef]

28. Wang, H.; Li, H.; Wang, J.; Jin, H. Reliability and Concurrent Validity of a Chinese Version of the Alberta Infant Motor Scale Administered to High-Risk Infants in China. *BioMed Res. Int.* **2018**, *2018*, 2197163. [CrossRef]

29. Mendoza-Pulido, C.; Ortiz-Corredor, F. Measurement properties of the Boston Carpal Tunnel Questionnaire in subjects with neurophysiological confirmation of carpal tunnel syndrome: A Rasch analysis perspective. *Qual. Life Res.* **2021**, *30*, 2697–2710. [CrossRef]

30. Mondelli, M.; Aprile, I.; Ballerini, M.; Ginanneschi, F.; Reale, F.; Romano, C.; Rossi, S.; Padua, L. Sex differences in carpal tunnel syndrome: Comparison of surgical and non-surgical populations. *Eur. J. Neurol.* **2005**, *12*, 976–983. [CrossRef]

31. Atroshi, I.; Lyrén, P.-E.; Gummesson, C. The 6-item CTS symptoms scale: A brief outcomes measure for carpal tunnel syndrome. *Qual. Life Res.* **2009**, *18*, 347–358. [CrossRef] [PubMed]

Systematic Review

Correlation between the Altered Gut Microbiome and Lifestyle Interventions in Chronic Widespread Pain Patients: A Systematic Review

María Elena Gonzalez-Alvarez [1,2], Eleuterio A. Sanchez-Romero [3,4,5,6,7,*], Silvia Turroni [8], Josué Fernandez-Carnero [1,2,3,5,6,7] and Jorge H. Villafañe [9,*]

1 Department of Physical Therapy, Occupational Therapy, Rehabilitation and Physical Medicine, Rey Juan Carlos University, 28032 Madrid, Spain
2 Escuela Internacional de Doctorado, Rey Juan Carlos University, 28008 Madrid, Spain
3 Department of Physiotherapy, Faculty of Sport Sciences, Universidad Europea de Madrid, 28670 Villaviciosa de Odón, Spain
4 Physiotherapy and Orofacial Pain Working Group, Sociedad Española de Disfunción Craneomandibular y Dolor Orofacial (SEDCYDO), 28009 Madrid, Spain
5 Musculoskeletal Pain and Motor Control Research Group, Faculty of Sport Sciences, Universidad Europea de Madrid, 28670 Villaviciosa de Odón, Spain
6 Department of Physiotherapy, Faculty of Health Sciences, Universidad Europea de Canarias, 38300 Santa Cruz de Tenerife, Spain
7 Musculoskeletal Pain and Motor Control Research Group, Faculty of Health Sciences, Universidad Europea de Canarias, 38300 Santa Cruz de Tenerife, Spain
8 Department of Pharmacy and Biotechnology, University of Bologna, Via Belmeloro 6, 40126 Bologna, Italy
9 IRCCS Fondazione Don Carlo Gnocchi, 20148 Milan, Italy
* Correspondence: eleuterio.sanchez@universidadeuropea.es (E.A.S.-R.); mail@villafane.it (J.H.V.)

Abstract: *Background*: Lifestyle interventions have a direct impact on the gut microbiome, changing its composition and functioning. This opens an innovative way for new therapeutic opportunities for chronic widespread patients. *Purpose*: The goal of the present study was to evaluate a correlation between lifestyle interventions and the gut microbiome in patients with chronic widespread pain (CWP). *Methods*: The systematic review was conducted until January 2023. Pain and microbiome were the two keywords selected for this revision. The search was conducted in PubMed, Chochrane, PEDro and ScienceDirect, where 3917 papers were obtained. Clinical trials with lifestyle intervention in CWP patients were selected. Furthermore, these papers had to be related with the gut microbiome, excluding articles related to other types of microbiomes. *Results*: Only six articles were selected under the eligibility criteria. Lifestyle interventions were exercise, electroacupuncture and ingesting a probiotic. *Conclusions*: Lifestyle intervention could be a suitable choice to improve the gut microbiome. This fact could be extrapolated into a better quality of life and lesser levels of pain.

Keywords: gut microbiome; lifestyle; chronic widespread pain

Citation: Gonzalez-Alvarez, M.E.; Sanchez-Romero, E.A.; Turroni, S.; Fernandez-Carnero, J.; Villafañe, J.H. Correlation between the Altered Gut Microbiome and Lifestyle Interventions in Chronic Widespread Pain Patients: A Systematic Review. *Medicina* **2023**, *59*, 256. https://doi.org/10.3390/medicina59020256

Academic Editors: Milica Lazovic and Dejan Nikolić

Received: 13 November 2022
Revised: 25 January 2023
Accepted: 26 January 2023
Published: 29 January 2023

1. Introduction

One in three people experience chronic pain, and at least one in ten experience an even greater symptomatology called chronic widespread pain (CWP), which carries a cost in absenteeism from work and represents a burden on the national health system and a worldwide problem [1–3]. CWP is a condition of diffuse musculoskeletal pain associated with other illness, which presented axial pain on both sides of the body [4]. The latest scientific literature has described the mechanisms underlying the communication between the gut and the brain (i.e., the gut–brain axis [5]), which could partly explain the chronicity of pain, as well as pave the way for new therapeutic opportunities for this population [2,6]. In particular, it is known that abnormalities in the gut microbiome (i.e., dysbiosis) could lead to systemic inflammation, especially in the presence of an impaired intestinal barrier (the so-called leaky gut) [7,8]. Moreover, some intestinal microorganisms can influence the production of neurotransmitters (and produce them by themselves) in addition to directly stimulating nerve fibers, also interfering in the hypothalamus–pituitary–adrenal axis [9]. Alterations along the gut–brain axis have been shown to be related with musculoskeletal pain, behavior modulation or brain processing, and play a role in depression, stress, anxiety and even neuropsychiatric disorders [10–13].

Lifestyle interventions such as nutrition, sleep or exercise could affect the pain experience [14–16]. Specifically, physical activity has a direct impact on the central nervous system (CNS), modifying the pain experience and cognitive processing [14,17]. On the other hand, it has been shown to positively modulate the gut microbiome (leading to greater diversity and overabundance of beneficial taxa and metabolites) in different settings, including chronic diseases [17,18] thus, representing a potential preventive and therapeutic tool for dysbiosis-related conditions.

As far as we are aware, the bibliography related to these three items (pain, microbiome and lifestyle) is still limited. Despite there being a wide bibliography about the effects of exercise in CWP patients, there is only one article in which exercise is the chosen intervention regardless of microbiota changes [18]. This stresses the need for further research in this very promising field, which could open a new vision and indicate new therapeutic targets for CWP patients.

Here, we provide an up-to-date systematic review of studies investigating the correlation between the gut microbiome and lifestyle interventions in CWP. In particular, the associations between gut microbiome and pain, quality of life and exercise are discussed.

2. Materials and Methods

2.1. Data Source and Search Strategy

A systematic review was conducted until January 2023 that investigated the correlation between the gut microbiome and lifestyle interventions in CWP. Preferred Reporting Items for Systematic Reviews and Meta-Analyses (PRISMA) guidelines were followed.

The criteria used to extract the data were according to the *Cochrane Handbook for Systematic Reviews of Interventions* (Version 6.3, 2022) [19]. The data selected were study design, age of participants, year and country of publication, setting, intervention, follow-up timing, clinical outcomes and reported findings. The protocol has been registered on 18 November 2022 in the International Prospective Register of Systematic Reviews (PROSPERO, CRD42022373890).

The principal researcher conducted the systematic review according to the PRISMA criteria by inserting the keywords "pain" and "microbiome", combined with the Boolean "AND" in PubMed, Cochrane, PEDro and ScienceDirect. This strategy was reviewed by two other authors. The whole search strategy used was: ("pain"[MeSH Terms] OR "pain"[All Fields]) AND ("microbiome"[All Fields] OR "microbiomic"[All Fields] OR "microbiomics"[All Fields] OR "microbiota"[MeSH Terms] OR "microbiota"[All Fields] OR "microbiome"[All Fields] OR "microbiomes"[All Fields]).

2.2. Eligibility Criteria

Eligibility criteria were: (1) clinical trial, (2) lifestyle intervention, (3) patients with CWP (4) in the gut microbiome and (5) until January 2023.

2.3. Data Extraction

Relevant articles were identified by the principal investigator (MEGA) and reviewed by two other authors (EASR and JHV). Discrepancies were resolved by the consensus of the three researchers. Researchers were not blind to any information regarding the authors, the journal or the outcomes for each article reviewed. The data extracted from the studies included study design, participants, intervention, outcome measures, follow-up and reported results (Table 1).

2.4. Outcome Measures

The primary outcome was the change in pain between baseline and follow-up. Furthermore, quality of life, scales related to their CWP symptoms and exercise were also added as secondary measures.

2.5. Quality Assessment

All articles selected by the principal author (MEGA) were assessed by two different authors (EASR and JHV), using two different scales for methodological quality. Control trial studies were evaluated with the Physiotherapy Evidence Database (PEDro) Scale (Table 2), while longitudinal studies were evaluated using the Methodological Index for Nonrandomized Studies (MINORS) (Table 3). Disagreements were solved by the three authors cited above. The entire quality assessment process was developed based on previous studies [20–23].

3. Results

3.1. Study Selection

A total of 3917 studies were identified by searching the PubMed, Cochrane, PEDro and ScienceDirect databases. No other studies were added from other sources. After removing the duplicates and screening titles, abstracts, and full texts, we selected eight articles. In a secondary screening, one from PubMed was discarded because the intervention was not completely related with our topic, and one from Cochrane was discarded because the registration of the Roman et al. [12] manuscript was included in the PubMed section. Six articles were therefore selected for this review, with 337 participants in total. Three studies were conducted in Europe (Spain, Italy and Denmark) and three in Asia (China and Turkey).

Although Roman et al. [12] and Jensen et al. [24] did not profile the gut microbiome, their works were included as they used probiotics and well-known microbiome manipulation tools [25]. The same scenery can be found in the Kenis-Coskun et al. [4] manuscript but in this case with Vitamin D [4]. Otherwise, Torlak et al. manipulated the diet and used different diets with chronic low back pain (CLBP) patients [26]. The other two studies investigated the effect of a maximal exercise challenge in myalgic encephalomyelitis/chronic fatigue syndrome (ME/CFS) [18] and the effect of electroacupuncture in knee osteoarthritis patients [27]. The flowchart is presented in Figure 1 according to the PRISMA guidelines.

3.2. Outcomes

3.2.1. Association between Gut Microbiome and Pain

The pain was measured through three different scales. Shukla et al. (2015) [18], Jensen et al. (2019) [24] and Wang et al. (2021) [27] used the numerical rating scale (NRS). Wang also used the Western Ontario and McMaster Universities Osteoarthritis Index (WOMAC) in the subscale of pain [27]. Roman et al. (2018) [12], Kenis-Coskun et al. (2020) [4] and Torlak et al. (2022) [26] chose the visual analog scale (VAS).

Shukla et al. (2015) [18] showed significant differences at baseline between the myalgic encephalomyelitis/chronic fatigue syndrome (ME/CFS) group and the control group in the relative abundance of the phylum Actinobacteria (higher in controls). It should be noted that ME/CFS patients experienced more pain (NRS mean: 6.8) than the control group (2.5).

Roman et al. (2018) [12] indicated a reduction in pain in the probiotic group compared to the placebo group but not statistically significant. Although they did not characterize the gut microbiome, this reduction is likely to be mediated by its modulation, as discussed by the authors (and suggested by the literature on the same probiotic strains used).

Table 1. Studies data extraction.

Author, Year.	Study Design	Participants	Intervention	Outcome Measures	Reported Results
Shukla et al., 2015 [18]	Randomized controlled trial	20 subjects (10 ME/CFS patients and 10 controls) Mean age of patients was 48.6 years Mean age of controls is 46.5 years Inclusion criteria: subjects with CFS without any other major illness	Maximal exercise test on an electronically braked cycle ergometerDuration of trial: 3-min warm-up at 25 W and the rate was increased to 5 W every 20 s. Participants should maintain a pedal rate between 60–70 RPM. The test ended if the participant could not maintain the pedal rate or stopped pedaling	Sample collection (blood and stool) DNA sequence analysis MFI POMS McGill Pain Questionnaire Symptom Inventory: Diarrhea and Stomach/Abdominal Pain Symptom Inventory: Neurocognitive Symptoms Memory ProblemsConcentration Problems Measured and Follow-up: Pre-intervention, immediately pre-test, immediately post-test, 40 h and 72 h post-test	Baseline: statistically significant differences between groups in MFI and POMS. A significant difference between groups in heart rate during the test. The relative abundance of Actinobacteria in the gut microbiome was significantly lower in ME/CFS patients than in healthy controls. Authors' conclusions: "Exercise induced bacterial translocation, one likely argument to why patients worsen when they try to be more physically active."
Roman et al., 2018 [12]	Randomized controlled trial	31 fibromyalgia participants Inclusion criteria: >1 year from the fibromyalgia diagnosed before the study. Exclusion criteria (1) use of antibiotics and nutritional supplements (2) allergies (3) participating in other psychological or medical studies (4) pregnant/ (5) breastfeeding (6) severe intestinal disease (7) psychiatric disorders	Probiotic ingests ($n = 16$) Ctrl: sham probiotics ($n = 15$) 2 pills before breakfast and dinner, for 8 weeks	VAS FIQ SF-36 STAI BDI MMSE Cognitive tasks (choice task and the Iowa gambling task) Cortisol Measures and Follow-up Pre- and post-intervention	Both groups: ↓ FIQ, depressive symptoms and cortisol ↑ SF-36 Authors' conclusions: "This intervention improves cognition, specifically impulsive choice and decision-making, in a group of patients diagnosed with fibromyalgia."

Table 1. *Cont.*

Author, Year.	Study Design	Participants	Intervention	Outcome Measures	Reported Results
Jensen et al., 2019 [24]	Randomized double-blind placebo controlled trial 1 year follow-up	85 participants, 42 active capsules and 43 placebo capsules. Inclusion criteria: CLBP and MC1 Age 18–65 Danish speakers RMQ < 5 Exclusion criteria: Intended, planned or previous back surgery Planned or current (last 3 months) antibiotic treatments for MC, immunosuppressants, intestinal pathology, immune deficiency, cancer or inability to complete de project	Active group: Probiotic Dicoflor ® twice daily for 100 days. Each capsule contains 6 billion Lactobacillus Rhamnosis GG. Placebo group: Placebo capsules indistinguishable from Dicoflor twice daily for 100 days	Age and sex Pain duration and intensity (NRS) Likert scale BMI RMQ Back+leg pain by the LBP rating scale Measures and Follow-up Until 1 year follow-up	No differences between intervention groups in regard to the predefined outcomes disability, back + leg pain, patient-reported global effect or the number of the patients with minimal disability at 1 year. Back pain decreased a little more in the active intervention group than in the control group. Authors' conclusions: "The study confirmed that treatment with probiotics, was safe and implicated no more side effects than placebo."
Kenis-Coskun et al., 2020 [4]	Clinical Trial	51 Female patients who have CWP with vitamin D deficiency Mean age: 44.3 ± 12.7 Mean symptom duration was 13.1 ± 6.7 months. Mean BMI: 21.6 ± 3.9 Exclusion criteria: Medications that alter CPS neuropathies Comorbidities that can cause vitamin D deficiency Rheumatologic or metabolic diseases Surgery (last 6 months) CNS disorders Psychiatric disorders BMI < 30.	8-week replacement therapy of vitamin D	VAS LANSS QoL (NHP) CSP parameters Vitamin D measurements Measures and Follow-up Before and after treatment	No significant changes un CSP parameters with Vitamin D replacement. Vitamin D replacement improve pain levels and QoL in patients with CWP. Authors' conclusions: "These results imply that in whichever way vitamin D is effective in CWP, it does not seem to be via the spinal inhibitory circuit that elicits the inhibition of muscle contraction with painful stimuli."

Table 1. *Cont.*

Author, Year.	Study Design	Participants	Intervention	Outcome Measures	Reported Results
Wang et al., 2021 [27]	Randomized controlled trial	90 participants (60 knee OA patients + 30 healthy controls) Inclusion criteria: Age 45–75 years, radiographically confirmed OA on one or both knees for more than 6 months and pain intensity ≥4 out of 10. Exclusion criteria: knee surgery, floating cartilage, joint effusion, inflammatory, malignant, or autoimmune disease, serious acute or chronic organic disease or mental disorder, pregnancy or breastfeeding, or history of bleeding disorder Participants were not included if they had acupuncture treatment or participated in other clinical trials in the past 3 months.	3 groups: EA: acupoints selected by formed acupuncturists ($n = 30$) SA: sham acupoints ($n = 30$) Healthy controls ($n = 30$) Duration of trial: 8 weeks	WOMAC (pain, stiffness and function subscales) NRS Function subscale: SF-12 Physical and mental health summary Fecal sample and DNA extraction 16S ribosomal RNA gene sequencing Microbial analysis Measures and Follow-up 0, 4, 8, 16, and 26 weeks of follow-up	↑ WOMAC total scores at 8 weeks in EA compared to SA ↑ WOMAC pain scores at 8 weeks in EA compared to SA ↓ NRS scores at 8 weeks in EA compared to SA Microbiota profiles were significantly different between EA (before intervention) and healthy controls. Blautia, Streptococcus and Eubacterium ↑ and Bacteroides and Agathobacter ↓ in EA. After treatment, Agathobacter and Lachnoclostridium ↑ EA. - Bacteroides were - correlated with NRS score, WOMAC total score, and WOMAC pain, stiffness and function. - - Agathobacter was - correlated with NRS score, WOMAC total score, and WOMAC pain, stiffness and function scores. - - Faecalibacterium was - correlated with NRS score, WOMAC total score, and WOMAC pain and function. - - Roseburia was - correlated with WOMAC total score, and WOMAC pain, stiffness and function. - Streptococcus was + correlated with NRS score, WOMAC total score, and WOMAC pain, stiffness and function scores. - - Enterococcus was + correlated with NRS score and WOMAC pain score. - Eubacterium, Blautia and Anaerostipes were positively correlated with SF-12 physiological score and SF-12 psychological score Authors' conclusions: - EA was more effective than SA at 8 weeks - EA treatment modified the diversity of the gut microbiome

Table 1. *Cont.*

Author, Year.	Study Design	Participants	Intervention	Outcome Measures	Reported Results
Torlak et al., 2022 [26]	Randomized controlled trial	60 sedentary, CLBP patients (30 female + 30 male) Inclusion criteria: VAS > 5 BMI > 25 kg/m^2. Exclusion criteria: Individuals who engage in active exercise Intake of painkillers, anti-depressant or cortisone Pregnant individuals; Severe chronic illness Spine surgery	PTG (n = 20) PT+DG (n = 20) DG (n = 20) DG lasted for 5 weeks, monitored daily PTG lasted for 5 weeks, monitored 5 times a week. Program: hot packs + US + TENS PT+DG used both intervention	BMI VAS LANSS BI Measures and Follow-up Before and after treatment	Significant difference in VAS scores in each group before and after treatment when intragroup values ere compared (PT+DG: $p < 0.001$; DG: $p < 0.001$; PTG: $p < 0.001$) No significant differences between groups in terms of VAS scores before and after treatment. Authors' conclusions: "Pain sensation decreased in all groups and the quality of life of the patients increased after treatment in PTG. Intermittent diet may be an alternative option for the treatment of chronic pain."

Abbreviations alphabetically ordered: ↑ = Increased; ↓ = Reduced; +: positively; -: negatively; BDI: Beck Depression Inventory; BI: Barthel Index; BMI: Body Mass Index; CFS: Chronic Fatigue Syndrome; CLBP: Chronic Low Back Pain; CPS: Central Pain Sensitivity; Ctrl: Control; DG: diet group; EA: Electroacupuncture; FIQ: Fibromyalgia Impact Questionnaire; LANSS: Leeds Assessment of Neuropathic Symptoms and Signs pain scale; MC1: Modic changes type 1; ME: Myalgic Encephalomyelitis; MFI: Multidimensional Fatigue Inventory; MMSE: Mini-Mental State Examination; NRS: Numerical rating scale for pain; NHP: Nottingham Health Profile; OA: Osteoarthritis; POMS: Profile of Mood States; PTG: Physical therapy group; PT+DG: Physical therapy + diet group; QoL: quality of Life; RMQ: Roland Morris Questionnaire; RPM: revolutions per minute; SA: Sham acupuncture; SF-12: The standard 12-item Short-Form Health Survey; SF-36: The SF-36 Quality of Life Questionnaire; STAI: 40-item State-Trait Anxiety Inventory; VAS: visual analog scale; W: watts; WOMAC: Western Ontario and McMaster Universities Osteoarthritis Index.

Table 2. Methodological quality evaluation of the clinical trials using the PEDro scale in randomized trials.

	Scale "Physiotherapy Evidence Database (PEDro)" to Analyze the Methodological Quality of Clinical Studies											
Authors	Specified Selection Criteria	Randomization	Allocation Was Concealed	Similar Groups to Start	Blinded Subjects	Blinded Therapists	Blinded Raters	Outcomes 85%	Treatment or Intention to Treat	Comparison between Groups	Point Measures Variability	Outcome
Roman et al., 2018 [12]	Yes	Yes	Yes	Yes	Yes	Yes	Yes	Yes	Yes	Yes	Yes	10
Wang et al., 2021 [27]	Yes	Yes *	No	Yes	No	No	No	Yes	Yes	Yes	Yes	7

Table 2. *Cont.*

		Scale "Physiotherapy Evidence Database (PEDro)" to Analyze the Methodological Quality of Clinical Studies										
Authors	Specified Selection Criteria	Randomization	Allocation Was Concealed	Similar Groups to Start	Blinded Subjects	Blinded Therapists	Blinded Raters	Outcomes 85%	Treatment or Intention to Treat	Comparison between Groups	Point Measures Variability	Outcome
Jensen et al., 2019 [24]	Yes	Yes	Yes	Yes	Yes	Yes	No	No	Yes	Yes	Yes	9
Kenis-Coskun et al., 2020 [4]	Yes	No	No	Yes	Yes	Yes	No	No	Yes	Yes	Yes	7
Torlak et al., 2022 [26]	Yes	Yes	Yes	Yes	Yes	No	No	Yes	Yes	Yes	Yes	9

Yes *: randomization between both groups of patients. There is a third non-randomized control group. Result on the PEDro scale: 9–10 (excellent), 6–8 (good), 4–5 (acceptable) and <4 (poor).

Table 3. Methodological index for non-randomized studies (MINORS) to assess the methodological quality and risk of bias of the included observational studies. Items are scored 0 (not reported), 1 (reported but inadequate) or 2 (reported and adequate), with the global ideal score being 16 for non-comparative studies and 24 for comparative studies.

		Methodological Index for Non-Randomized Studies (MINORS)										
Authors	A Clearly Stated Aim	Inclusion of Consecutive Patients	Prospective Collection of Data	Endpoints Appropriate to the Aim of the Study	Unbiased Assessment of the Study Endpoint	Follow-Up Period Appropriate to the Aim of the Study	Loss to Follow Up Less than 5%	Prospective Calculation of the Study Size	An Adequate Control Group *	Contemporary Groups *	Baseline Equivalence of Groups *	Adequate Statistical Analyses *	Outcome
Shukla et al., 2015 [18]	2	2	1	2	0	1	2	1	2	2	1	1	17

The items are scored 0 (not reported), 1 (reported but inadequate) or 2 (reported and adequate). The global ideal score being 16 for non-comparative studies and 24 for comparative studies. * Additional criteria in case of comparative study.

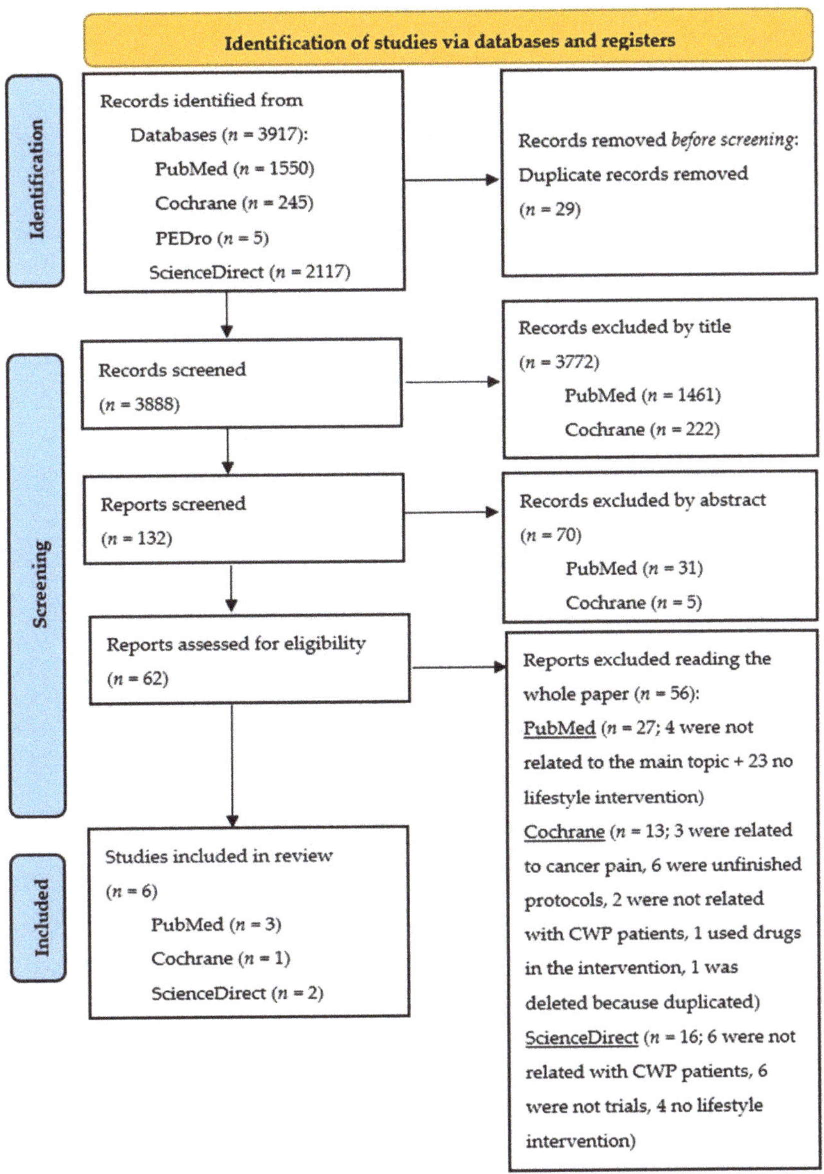

Figure 1. PRISMA 2020 flow diagram for new systematic reviews, including searches of databases and registers.

Jensen et al. (2019) [24] and Kenis-Coskun et al. (2020) [4] found improvement in pain after treatment; however, the differences were not significant. In particular, Jensen [24] did not show any differences between the control and active group in four out of five items: disability, back+leg pain, patient-reported global effect and number of the patients with minimal disability at 1 year. Nevertheless, authors highlighted that back pain decreased more in the active group than in the control group. Kenis-Coskum et al. [4] showed a decrease in the VAS median from 7/10 before treatment to 3/30 after treatment [4].

Wang et al. (2021) [27] mentioned a statistical difference between cases and controls at week 8 in NRS and WOMAC pain (i.e., lower values for cases). They also showed an overall reduction in pain during follow-up through week 26. The authors correlated this decrease with increased proportions of typically health-associated taxa, such as Faecalibacterium, Roseburia and Agathobacter (for WOMAC pain only). On the other hand, Streptococcus and Enterococcus, known pathobionts, were positively correlated with NRS and WOMAC pain.

Finally, Torlak et al. (2022) [26] found significant differences in VAS scores in intragroup comparing before and after treatment, but, in general terms before and after the treatment, there were not significant differences. As with Kenis-Coskum et al. [4], authors express the VAS in cm being the most important data for the VAS in the diet and physical therapy group before treatment 7.45 ± 0.44 and after treatment 4.7 ± 0.42 (<0.001).

3.2.2. Association between the Gut Microbiome and Quality of Life

Roman et al. (2018) [12] and Wang et al. (2021) [27] measured quality of life with the SF-36 questionnaire. In particular, Wang et al. (2021) [27] used the abbreviated form, SF-12, which yields physical and mental health. According to their findings, both items improved during the follow-up. As for the gut microbiome, Blautia and Anaerostipes were positively correlated with SF-12, in the physiological and psychological scores. On the other hand, Roman et al. (2018) [12] showed better results in the SF-36 test for both the probiotic and placebo groups. No data on quality of life were provided by Shulka et al. (2015) [18].

Kenis-Coskun et al. (2020) [4] measured the quality of life through the Nottingham Health Profile (NHP) scale. This scale showed enhancements in five out of six items: energy, pain, emotional, sleep and physical being, with social life the only item where there were no improvements.

Torlak et al. (2022) [26] used the Barthel Index (BI) as a measure of quality of life using basic daily life activities. The disability in diet+physical therapy group and physical therapy group decreased significantly before and after treatment but not in the diet group.

3.2.3. Association between the Gut Microbiome and the Exercise

There were several changes in the gut microbiome in all study participants after a maximal exercise challenge (Shukla et al., 2015 [18]). In the ME/CFS group, six out of the nine major taxa (mainly Clostridium cluster IV, clostridia, bacilli, Firmicutes and Actinobacteria) increased in stool from baseline at 72 h post-exercise compared with only two (Bacteroidetes and unclassified general) in the control group. The authors consistently observed rapid changes (i.e., increases) in Firmicutes levels in blood samples 15 min after maximal exercise, a phenomenon that might be more evident in patients than controls.

4. Discussion

The purpose of this systematic review was to summarize available evidence from clinical trials, linking lifestyle intervention and gut microbiome in CWP patients, a field still underexplored despite its promising potential. Although no filter was applied by the year of publication, only six trials had the minimum requirements to be included. Their results hypothesized favorable changes in gut microbiome composition due to lifestyle intervention (exercise vs. electroacupuncture vs. probiotics vs. vitamin D vs. diet changes). Overall, these changes were associated with improved pain experience and quality of life. Despite the different levels of taxonomic resolution among studies, microbiome variations mainly involved an increase in beneficial taxa, such as those producing short-chain fatty acids (mainly acetate, propionate and butyrate), which are microbial metabolites with a key multifactorial role in host physiology [28]. Despite the literature reporting some conflicting data [28,29], it is known that short-chain fatty acids have a general anti-inflammatory effect as well as immunoregulatory and neuromodulatory effects, acting at different levels along the gut–brain axis, thus potentially contributing to pain relief [30].

Regarding specific lifestyle interventions, a wide range of literature supports the therapeutic effects of exercise on pain intensity in CWP [31,32]. Recent bibliography also claims that exercise could modify the gut microbiome, with compositional and functional changes likely related to training variables, such as type, load, intensity, frequency, etc. [31,33,34]. In this regard, it should be noted that Shukla et al. (2015) [18] found evidence of bacterial translocation into the bloodstream after a maximal exercise challenge, thus suggesting that too intense exercise may not be entirely favorable.

On the other hand, probiotics are well-known microbiome manipulation tools with a long history of use [25]. However, both Roman et al. (2018) [12] and Jensen et al. (2019) [24] did not profile the gut microbiome and observed improvements even in the placebo group.

Treatment with vitamin D could be a suitable ally in the fight against CWP. Nevertheless, clinicians should take care regarding the quantity of vitamin D, as other authors reported that high levels of vitamin D could increase the pro-inflammatory mediators and, consequentially, increase the levels of pain [35].

Finally, it is hypothesized that electroacupuncture could slightly modify pain in knee OA because of its relationship with the inflammatory effects, reducing the quantity of Streptococcus. However, this association is still not clear enough [27]. Other authors also suggest the same hypothesis, however, manipulating directly the microbiome for modifying the knee OA pain [36].

CWP patients suffer from malfunctioning sensory processing in the CNS [37]. This may lead to deficiencies in the pain processing chain, such as descending pain-inhibitory mechanisms or temporal summation (TS) [38]. Not only functional changes, but the literature also explains structural neuroplastic changes related to this sensibilization [39]. Those changes could be promoted by the alterations in the gut microbiome, increasing the inflammatory response and chronification of the illness [5,36,40]. Therefore, there is a necessity of thinking about interventions that make an impact on the musculoskeletal, the microbiome and, as consequence, in the CNS [6,40,41].

It is complicated to draw firm conclusions about changes in pain sensitivity and gut microbiome solely due to the interventions mentioned before. For example, diet is a major microbiome-associated cofounding factor that, in many trials, has not been taken into account [34]. Additional research is therefore required to determine the best intervention to change microbiome composition and functionality towards a eubiotic profile capable of counteracting (i.e., decreasing) CWP. Such research should also provide insights into the underlying mechanisms, to be possibly validated in animal models.

It is important to highlight that despite all the interventions that have an impact in the microbiome, only three studies carried out a specific measurement of it. This fact leads us to explain that, in those where there is no analysis of the microbiome, its association is a hypothesis that will have to be contrasted with more scientific studies in this field.

This review has several limitations. One of them is the differences between protocols in order to measure the impact of each intervention on the microbiome. Although all the patients suffer from CWP, they were diagnosed with diverse illnesses. Appreciating those differences as well as the differences in the methodologies make this article out from the definition of a meta-analysis.

5. Conclusions

Lifestyle is one of the major variation drivers of the gut microbiome. Given the latter's role in human pathophysiology, lifestyle interventions could result in microbiome-related improvements in pain and quality of life in several chronic widespread diseases.

Author Contributions: Conception and design, M.E.G.-A., J.H.V. and E.A.S.-R.; analysis and interpretation of the data, M.E.G.-A., J.H.V. and E.A.S.-R.; writing—original draft preparation, M.E.G.-A., J.H.V., E.A.S.-R. and S.T.; writing—review and editing, M.E.G.-A., J.H.V., E.A.S.-R., S.T. and J.F.-C.; visualization, M.E.G.-A., J.H.V., E.A.S.-R., S.T. and J.F.-C.; supervision, all authors. E.A.S.-R. oversaw the review process and managed article processing charges. All authors have read and agreed to the published version of the manuscript.

Funding: This research received no external funding.

Institutional Review Board Statement: Not applicable.

Informed Consent Statement: Not applicable.

Data Availability Statement: The data presented in this study are available on request from the corresponding authors.

Acknowledgments: This study was supported and funded by the Italian Ministry of Health—Ricerca Corrente, 2022.

Conflicts of Interest: The authors declare no conflict of interest.

References

1. Andrews, P.; Steultjens, M.; Riskowski, J. Chronic widespread pain prevalence in the general population: A systematic review. *Eur. J. Pain* **2018**, *22*, 5–18. [CrossRef] [PubMed]
2. Freidin, M.B.; Stalteri, M.A.; Wells, P.M.; Lachance, G.; Baleanu, A.-F.; Bowyer, R.C.E.; Kurilshikov, A.; Zhernakova, A.; Steves, C.J.; Williams, F.M.K. An association between chronic widespread pain and the gut microbiome. *Rheumatology* **2021**, *60*, 3727–3737. [CrossRef]
3. Nijs, J.; Tumkaya Yilmaz, S.; Elma, Ö.; Tatta, J.; Mullie, P.; Vanderweeën, L.; Clarys, P.; Deliens, T.; Coppieters, I.; Weltens, N.; et al. Nutritional intervention in chronic pain: An innovative way of targeting central nervous system sensitization? *Expert Opin. Ther. Targets* **2020**, *24*, 793–803. [CrossRef]
4. Kenis-Coskun, O.; Giray, E.; Gunduz, O.H.; Akyuz, G. The effect of vitamin D replacement on spinal inhibitory pathways in women with chronic widespread pain. *J. Steroid Biochem. Mol. Biol.* **2020**, *196*, 105488. [CrossRef] [PubMed]
5. Villafañe, J.H.; Drago, L. What is the site of pain osteoarthritis? A triple gut-brain-joint microbioma axis. *Ann. Rheum. Dis.* **2019**, *37*, 20–21.
6. Pedersini, P.; Turroni, S.; Villafañe, J.H. Gut microbiota and physical activity: Is there an evidence-based link? *Sci. Total. Environ.* **2020**, *727*, 138648. [CrossRef]
7. Boer, C.G.; Radjabzadeh, D.; Medina-Gomez, C.; Garmaeva, S.; Schiphof, D.; Arp, P.; Koet, T.; Kurilshikov, A.; Fu, J.; Ikram, M.A.; et al. Intestinal microbiome composition and its relation to joint pain and inflammation. *Nat. Commun.* **2019**, *10*, 4881. [CrossRef] [PubMed]
8. Villafañe, J.H. Preface. *Top. Geriatr. Rehabil.* **2021**, *37*, 207–208. [CrossRef]
9. Mitrea, L.; Nemes, S.A.; Szabo, K.; Teleky, B.E.; Vodnar, D.C. Guts Imbalance Imbalances the Brain: A Review of Gut Microbiota Association with Neurological and Psychiatric Disorders. *Front. Med.* **2022**, *9*, 813204. [CrossRef]
10. Mayer, E.A.; Tillisch, K.; Gupta, A. Gut/brain axis and the microbiota. *J. Clin. Investig.* **2015**, *125*, 926–938. [CrossRef]
11. Naveed, M.; Zhou, Q.-G.; Xu, C.; Taleb, A.; Meng, F.; Ahmed, B.; Zhang, Y.; Fukunaga, K.; Han, F. Gut-brain axis: A matter of concern in neuropsychiatric disorders . . . ! *Prog. Neuro-Psychopharmacol. Biol. Psychiatry* **2021**, *104*, 110051. [CrossRef] [PubMed]
12. Roman, P.; Estévez, A.F.; Miras, A.; Sánchez-Labraca, N.; Cañadas, F.; Vivas, A.B.; Cardona, D. A Pilot Randomized Controlled Trial to Explore Cognitive and Emotional Effects of Probiotics in Fibromyalgia. *Sci. Rep.* **2018**, *8*, 10965. [CrossRef] [PubMed]
13. Almagro, J.R.; Almagro, D.R.; Ruiz, C.S.; González, J.S.; Martínez, A.H. The Experience of Living with a Gluten-Free Diet. *Gastroenterol. Nurs.* **2018**, *41*, 189–200. [CrossRef] [PubMed]
14. Borisovskaya, A.; Chmelik, E.; Karnik, A. Exercise and Chronic Pain. *Adv. Exp. Med. Biol.* **2020**, *1228*, 233–253. [CrossRef] [PubMed]
15. Elma, Ö.; Yilmaz, S.T.; Deliens, T.; Coppieters, I.; Clarys, P.; Nijs, J.; Malfliet, A. Nutritional factors in chronic musculoskeletal pain: Unravelling the underlying mechanisms. *Br. J. Anaesth.* **2020**, *125*, e231–e233. [CrossRef] [PubMed]
16. Villafañe, J.H.; Pirali, C.; Dughi, S.; Testa, A.; Manno, S.; Bishop, M.D.; Negrini, S. Association between malnutrition and Barthel Index in a cohort of hospitalized older adults article information. *J. Phys. Ther. Sci.* **2016**, *28*, 607–612. [CrossRef]
17. Negrini, S.; Imperio, G.; Villafañe, J.H.; Negrini, F.; Zaina, F. Systematic reviews of physical and rehabilitation medicine Cochrane contents. Part 1. Disabilities due to spinal disorders and pain syndromes in adults. *Eur. J. Phys. Rehabil. Med.* **2013**, *49*, 595–596.
18. Shukla, S.K.; Cook, D.; Meyer, J.; Vernon, S.D.; Le, T.; Clevidence, D.; Robertson, C.E.; Schrodi, S.; Yale, S.; Frank, D.N. Changes in Gut and Plasma Microbiome following Exercise Challenge in Myalgic Encephalomyelitis/Chronic Fatigue Syndrome (ME/CFS). *PLoS ONE* **2015**, *10*, e0145453. [CrossRef]
19. Higgins, J.P.T.; Thomas, J.; Chandler, J.; Cumpston, M.; Li T.; Page, M.J.; Welch, V.A. (Eds.) *Cochrane Handbook for Systematic Reviews of Interventions Version 6.3*; (Updated February 2022); Cochrane: London, UK, 2022. Available online: www.training.cochrane.org/handbook (accessed on 25 January 2023).
20. Meléndez Oliva, E.; Villafañe, J.H.; Alonso Pérez, J.L.; Alonso Sal, A.; Molinero Carlier, G.; Quevedo García, A.; Turroni, S.; Martínez-Pozas, O.; Valcárcel Izquierdo, N.; Sánchez Romero, E.A. Effect of Exercise on Inflammation in Hemodialysis Patients: A Systematic Review. *J. Pers. Med.* **2022**, *12*, 1188. [CrossRef]

21. Sánchez Romero, E.A.; Martínez-Pozas, O.; García-González, M.; de-Pedro, M.; González-Álvarez, M.E.; Esteban-González, P.; Cid-Verdejo, R.; Villafañe, J.H. Association between Sleep Disorders and Sleep Quality in Patients with Temporomandibular Joint Osteoarthritis: A Systematic Review. *Biomedicines* **2022**, *10*, 2143. [CrossRef]

22. Schulté, B.; Nieborak, L.; Leclercq, F.; Villafañe, J.H.; Sánchez Romero, E.A.; Corbellini, C. The Comparison of High-Intensity Interval Training Versus Moderate-Intensity Continuous Training after Coronary Artery Bypass Graft: A Systematic Review of Recent Studies. *J. Cardiovasc. Dev. Dis.* **2022**, *9*, 328. [CrossRef]

23. Reinert, G.; Müller, D.; Wagner, P.; Martínez-Pozas, O.; Cuenca-Zaldivar, J.N.; Fernández-Carnero, J.; Sánchez Romero, E.A.; Corbellini, C. Pulmonary Rehabilitation in SARS-CoV-2: A Systematic Review and Meta-Analysis of Post-Acute Patients. *Diagnostics* **2022**, *12*, 3032. [CrossRef] [PubMed]

24. Jensen, O.K.; Andersen, M.H.; Østgård, R.D.; Andersen, N.T.; Rolving, N. Probiotics for chronic low back pain with type 1 Modic changes: A randomized double-blind, placebo-controlled trial with 1-year follow-up using Lactobacillus Rhamnosis GG. *Eur. Spine J.* **2019**, *28*, 2478–2486. [CrossRef] [PubMed]

25. Suez, J.; Zmora, N.; Segal, E.; Elinav, E. The pros, cons, and many unknowns of probiotics. *Nat. Med.* **2019**, *25*, 716–729. [CrossRef]

26. Torlak, M.S.; Bagcaci, S.; Akpinar, E.; Okutan, O.; Nazli, M.S.; Kuccukturk, S. The effect of intermittent diet and/or physical therapy in patients with chronic low back pain: A single-blinded randomized controlled trial. *Explore* **2022**, *18*, 76–81. [CrossRef] [PubMed]

27. Wang, T.-Q.; Li, L.-R.; Tan, C.-X.; Yang, J.-W.; Shi, G.-X.; Wang, L.-Q.; Hu, H.; Liu, Z.-S.; Wang, J.; Wang, T.; et al. Effect of Electroacupuncture on Gut Microbiota in Participants with Knee Osteoarthritis. *Front. Cell. Infect. Microbiol.* **2021**, *11*, 597431. [CrossRef] [PubMed]

28. Koh, A.; De Vadder, F.; Kovatcheva-Datchary, P.; Bäckhed, F. From Dietary Fiber to Host Physiology: Short-Chain Fatty Acids as Key Bacterial Metabolites. *Cell* **2016**, *165*, 1332–1345. [CrossRef] [PubMed]

29. Lanza, M.; Filippone, A.; Ardizzone, A.; Casili, G.; Paterniti, I.; Esposito, E.; Campolo, M. SCFA Treatment Alleviates Pathological Signs of Migraine and Related Intestinal Alterations in a Mouse Model of NTG-Induced Migraine. *Cells* **2021**, *10*, 2756. [CrossRef]

30. Li, S.; Hua, D.; Wang, Q.; Yang, L.; Wang, X.; Luo, A.; Yang, C. The Role of Bacteria and Its Derived Metabolites in Chronic Pain and Depression: Recent Findings and Research Progress. *Int. J. Neuropsychopharmacol.* **2020**, *23*, 26–41. [CrossRef]

31. Villafañe, J.H.; Bishop, M.D.; Pedersini, P.; Berjano, P. Physical Activity and Osteoarthritis: Update and Perspectives. *Pain Med.* **2019**, *20*, 1461–1463. [CrossRef]

32. Sánchez, E.; Fernandez-Carnero, J.; Calvo-Lobo, C.; Sáez, V.O.; Caballero, V.B.; Pecos-Martín, D. Is a Combination of Exercise and Dry Needling Effective for Knee OA? *Pain Med.* **2020**, *21*, 349–363. [CrossRef]

33. Villafañe, J.H. Exercise and osteoarthritis: An update. *J. Exerc. Rehabil.* **2018**, *14*, 538–539. [CrossRef]

34. Mailing, L.J.; Allen, J.M.; Buford, T.W.; Fields, C.J.; Woods, J.A. Exercise and the Gut Microbiome: A Review of the Evidence, Potential Mechanisms, and Implications for Human Health. *Exerc. Sport Sci. Rev.* **2019**, *47*, 75–85. [CrossRef]

35. Gominak, S. Vitamin D deficiency changes the intestinal microbiome reducing B vitamin production in the gut. The resulting lack of pantothenic acid adversely affects the immune system, producing a "pro-inflammatory" state associated with atherosclerosis and autoimmunity. *Med. Hypotheses* **2016**, *94*, 103–107. [CrossRef]

36. Pedersini, P.; Savoldi, M.; Berjano, P.; Villafañe, J.H. A probiotic intervention on pain hypersensitivity and microbiota composition in patients with osteoarthritis pain: Study protocol for a randomized controlled trial. *Arch. Rheumatol.* **2021**, *36*, 296–301. [CrossRef]

37. Villafañe, J.H.; Valdes, K.; Pedersini, P.; Berjano, P. Osteoarthritis: A call for research on central pain mechanism and personalized prevention strategies. *Clin. Rheumatol.* **2019**, *38*, 583–584. [CrossRef]

38. Villafañe, J.H. Does "time heal all wounds" still have a future in osteoarthritis? *Clin. Exp. Rheumatol.* **2018**, *36*, 513.

39. Pedersini, P.; Gobbo, M.; Bishop, M.D.; Arendt-Nielsen, L.; Villafañe, J.H. Functional and Structural Neuroplastic Changes Related to Sensitization Proxies in Patients with Osteoarthritis: A Systematic Review. *Pain Med.* **2022**, *23*, 488–498. [CrossRef]

40. Drago, L.; Zuccotti, G.V.; Romanò, C.L.; Goswami, K.; Villafañe, J.H.; Mattina, R.; Parvizi, J. Oral–Gut Microbiota and Arthritis: Is There an Evidence-Based Axis? *J. Clin. Med.* **2019**, *8*, 1753. [CrossRef]

41. Sánchez Romero, E.A.; Meléndez Oliva, E.; Alonso Pérez, J.L.; Martín Pérez, S.; Turroni, S.; Marchese, L.; Villafañe, J.H. Rela-tionship between the Gut Microbiome and Osteoarthritis Pain: Review of the Literature. *Nutrients* **2021**, *13*, 716. [CrossRef]

MDPI

St. Alban-Anlage 66

4052 Basel

Switzerland

www.mdpi.com

Medicina Editorial Office

E-mail: medicina@mdpi.com

www.mdpi.com/journal/medicina